LIST

TECHNOLOGY AND EMPIRE
GEORGE GRANT

LIST

First published in 1969 by House of Anansi Press Inc. This edition published
in Canada in 2018 and the USA in 2018 by House of Anansi Press Inc.
www.houseofanansi.com

House of Anansi Press is committed to protecting our natural environment.
As part of our efforts, the interior of this book is printed on paper that
contains 100% post-consumer recycled fibres, is acid-free, and is processed
chlorine-free.

22 21 20 19 18   1 2 3 4 5

Library and Archives Canada Cataloguing in Publication

Grant, George, 1918–1988, author
Technology and empire / George Grant.

ISBN 978-1-4870-0457-6 (softcover)

1. Technology—Philosophy. 2. Technology and civilization.
3. North America—Civilization. 4. Canada—Relations—United States.
5. United States—Relations—Canada. 6. Political science. I. Title.

E40.G7 2018   970   C2018-900689-7

Library of Congress Control Number: 2018940095

"Religion and State" appeared in *Queen's Quarterly* in 1963. "Tyranny
and Wisdom" was first published in *Social Research* in 1964. *Canadian
Dimension* printed "Canadian Fate and Imperialism" in 1967; its present form
is a revision of that paper. "The University Curriculum" appeared in *This
Magazine Is About Schools* (1967–1968) and in *The University Game* (1968); it
too has been revised. "In Defense of North America" was written in 1968. "A
Platitude" was written in 1968–1969, and has appeared in *Saturday Night*.

Series design: Brian Morgan
Cover illustration: Patrick Gray
Typesetting: Sara Loos

*We acknowledge for their financial support of our publishing program the
Canada Council for the Arts, the Ontario Arts Council, and the Government
of Canada.*

Printed and bound in Canada

RECYCLED
Paper made from
recycled material
FSC® C103567
www.fsc.org

INTRODUCTION
by Andrew Potter

GEORGE GRANT IS MOSTLY REMEMBERED today for *Lament for a Nation*, his by turns fiery and mournful 1965 pamphlet on the failure of English-speaking Canadian nationalism. Writing in the aftermath of the victory of Lester B. Pearson's Liberals over John Diefenbaker in the 1963 federal election, Grant saw the defeat of the Conservatives as marking the end of the possibility of Canada as a society distinct from, and independent of, the United States.

He was clearly mistaken. If anything, Canadian nationalism is stronger today than it was in Grant's time, and Canada remains a sovereign entity. It is a different nationalism than the one he envisaged, and our sovereignty has certainly evolved over time, but it cannot be denied that on the key points about Canada's political future, Grant was far too pessimistic.

So why read him today? What can be learned by returning to the ideas and arguments of a man who, looking back, seems so completely embedded in his time? The answer is that whatever his shortcomings as a pundit, as a political philosopher Grant understood as well as anyone the existential crisis facing humanity in the technological age.

As Grant saw it, Diefenbaker's defeat was just the localized conse-quence of a deeper quandary, namely, Canada's position as an adjunct to the global centre of technological liberalism, the United States of America. What it means to live inside a technological society is the recurring question motivating the collection of essays in *Technology and Empire*, the more academic work that followed the remarkable success of *Lament*. In Grant's view, to be a North American is to be cut off from our civilization's two great sources of meaning: Greece and Jerusalem, philosophy and religion. What these have in common is that they both provide complete theories of the good, giving a unified and comprehensive moral, political, and aesthetic account of human flourishing, whether it is achieved through contemplation (Greece) or revelation (Jerusalem).

In contrast, North America is the centre of a new civilization that privileges a liberal ideology according to which humanity's essence lies in the freedom to remake the world as we desire, employing the ever more efficient tools of technology. One consequence of the way liberalism conceives of freedom is that it closes us off from any comprehensive and publicly shared conception of the good. Each of us is free to pursue our own good according to our chosen values, as long as this pursuit is consistent with an equal amount of freedom for everyone else.

Why should we resist the emergence of a liberal society in which the good is relegated to the private realm? The argument for this stance is developed here in what is perhaps Grant's most important scholarly paper, "Tyranny and Wisdom." In this essay, Grant mediates a dispute between Alexandre Kojève and Leo Strauss on the question of whether the emergence of a universal and homogeneous state would be a tyranny, or mark the end of human striving and struggle. Essentially a continuation of an ongoing dispute between Hegel and the Greeks, the upshot is that Grant sides with Strauss against Kojève in holding that the universal and homogeneous state would be a tyranny. Grant's contribution to the debate is to propose that this universal and homogeneous state will emerge first under the guise of a liberal-technological-consumerist empire spearheaded by the United States of America.

There is a gloomy mood hanging over a lot of the writing here, with gusts to the apocalyptic. "What is worth doing in this barren twilight," he writes at the end of one essay, "is the incredibly difficult question."

In another, he suggests that "the sheer aridity of the public world will indeed drive many to seek excellence in strange and dangerous kingdoms (as those of drugs and myth and sexuality)."

At times, Grant seems overwhelmed by the sense of what has been lost in being North American, and shockingly unconcerned with the reasons why the old systems of meaning failed. And so even as he accepts that the drive for moral perfectionism was shot through with a host of "illusions, horizons, superstitions, taboos" that for centuries or millennia served to oppress, enslave, and bind entire peoples, he ties himself into knots defending the systems that underwrote this oppression. "This may be the case. What we lost may have been bad for men. But this does not change the fact that something has been lost."

More troubling is how the fact of deep pluralism with respect to values appears to mystify him, and he has trouble saying anything remotely positive about the modern world. Liberalism strikes him as profoundly shallow and pointless, which leads him to treat deeply human drives as diverse as sexual orientation and spiritual thirst as equally shallow expressions of mere consumerism: "Some like pizza, some like steaks; some like girls, some like boys; some like synagogues, some like the Mass. But we all do it in churches, motels, restaurants indistinguishable from the Atlantic to the Pacific."

For all this, bursts of insight cut through the gloom like stabs of sunshine. In the essay "Canadian Fate and Imperialism," Grant describes the Vietnam War as a crime committed "by the English-speaking empire and in the name of liberal democracy." And as he rightly points out, the fact that Canada did not send troops to Vietnam does not alter the fact that Canada was deeply implicated in that war not just economically and politically, but existentially. Grant summarizes it all by saying, "One is tempted to state that the North American motto is 'the orgasm at home and napalm abroad.'" Adapted to more recent events, Grant's arguments serve as a useful corrective to the ongoing Canadian smugness at having "stayed out" of the 2003 invasion of Iraq.

George Grant liked to describe the technological civilization as having reached "the end of ideology," but thanks to Francis Fukuyama's famous book on the subject, we now describe it as "the end of history." Both men looked askance at the liberal project and worried about what its end game might look like. Fukuyama foresaw two possibilities:

a retreat into cultural nostalgia on the one hand, and the pursuit of an ecstatic posthumanism on the other.

For Grant, the final stage of the liberal discourse is nihilism. We reason instrumentally about the pursuit of ends that are given, but where do we discover our ends? Where do our purposes come from? As much as liberals would like to wish the problem away, the question of the source of ultimate meaning in an indifferent universe remains unanswered. For George Grant, the two paths out of this spiritual dead end are nostalgia and madness. Were Grant alive in 2018, he would be utterly unsurprised to see these two paths joined in one terrifying political movement.

ANDREW POTTER is a journalist and academic. He holds a Ph.D. in philosophy from the University of Toronto. From 2006 to 2011 he was a public affairs columnist for *Maclean's* magazine. Between 2007 and 2015 he was an editor at the *Ottawa Citizen*, serving as the editor-in-chief from 2013 to 2015. He is now an Associate Professor at McGill University, where his research and teaching focus on the intersection of technology, politics, and media.

TECHNOLOGY AND EMPIRE

For S.V.G. and D.B.L.

sine qua non

CONTENTS

PREFACE

These essays are published together because they are all perspectives on what it is to live in the Great Lakes region of North America. They do not presume to be philosophy, but are written out of the study of the history of political philosophy. If they seem too austere, I would say that they were mostly written as the meaning of the English-speaking world's part in the Vietnam war gradually presented its gorgon's face. How could there be any public laughter for somebody whose life came forth from the English-speaking world, at a time when that world reached its basest point? The purpose of the art of comedy is to bring together justice and felicity. In the face of being party to that outrage one cannot hope to attempt that uniting.

The reader's indulgence is required because the way I use certain key words is often made clear in one essay and then assumed elsewhere. For example, what I mean by "liberalism" (as a modern phenomenon) is explicitly defined only in the article on the curriculum (page 102). The phrase "the universal and homogeneous state" is clarified in the essay "Tyranny and Wisdom" (page 77) and used elsewhere in the light of that. The reading is also complicated by the fact that the essays are not arranged in the order they were written. Thus Professor Ellul's definition of "technique" is quoted in the article on the curriculum (page 101), while my criticism of that definition is implied throughout "In Defence of North America," the first essay in the book.

G. P. Grant
Dundas, 1969

In Defence
of
North America

TO EXIST AS A NORTH AMERICAN is an amazing and enthralling fate. As in every historical condition, some not only have to live their fate, but also to let it come to be thought. What we have built and become in so short a time calls forth amazement in the face of its novelty, an amazement which leads to that thinking. Yet the very dynamism of the novelty enthralls us to inhibit that thinking.

It is not necessary to take sides in the argument between the ancients and moderns as to what is novelty, to recognize that we live in novelty of some kind. Western technical achievement has shaped a different civilization from any previous, and we North Americans are the most advanced in that achievement. This achievement is not something simply external to us, as so many people envision it. It is not merely an external environment which we make and choose to use as we want — a playground in which we are able to do more and more, an orchard where we can always pick variegated fruit. It moulds us in what we are, not only at the heart of our animality in the propagation and continuance of our species, but in our actions and thoughts and imaginings. Its pursuit has become our dominant activity and that dominance fashions both the public and private realms. Through that achievement we have become the heartland of the wealthiest and most powerful empire that has yet been. We can exert our influence over a greater extent of the globe and take a greater tribute of wealth than any previously. Despite our limitations and miscalculations, we have more compelling means than any previous for putting the brand of our civilization deeply into the flesh of others.

To have become so quickly the imperial centre of an increasingly realized technological civilization would be bewildering for any human beings, but for North Americans particularly so. From our beginnings there has been an ambiguity for us as to who we are. To the Asians as they suffer from us, we must appear the latest wave of dominating Europeans who spread their ways around the world, claiming that those ways were not simply another civilization, but the highest so far, and whose claim was justified in the fact of power, namely that it could only be countered by Asians who accepted the very forms which threatened them. To the Europeans also we appear as spawned by themselves: the children of some low class servants who once dared to leave the household and who now surprisingly appear as powerful and dominating neighbours masquerading as gentry, whose threat can only be minimized by teaching them a little culture. They express contempt of us as a society barren of anything but the drive to technology; yet their contempt is too obviously permeated with envy to be taken as pure.

In one sense both the Asians and Europeans are correct. Except for the community of the children of the slaves and the few Indians we have allowed just to survive, we are indeed Europeans. Imperially we turn out to the rest of the world bringing the apogee of what Europeans first invented, technological civilization. Our first ways, in terms of which we met the new land, came with us from Europe and we have always used our continuing contact with the unfolding of that civilization. To this day many of our shallow intellectual streams are kept flowing by their rain. It was exiled Europeans with the new physical theory who provided us with our first uses of atomic energy. Our new social science may fit us so perfectly as to seem indigenous; but behind Parsons is Weber, behind Skinner, Pavlov, behind social work and psychiatry Freud. Even in seeking some hope against the inhuman imperial system and some less sterile ground of political morality than a liberalism become the end of ideology, many of the most beautiful young turn for their humanism to so European a thinker as Marcuse. In a field as un-American as theology, the continually changing ripples of thought, by which the professionals hope to revive a dying faith, originate from some stone dropped by a European thinker.

Yet those who know themselves to be North Americans know they are not Europeans. The platitude cannot be too often stated that the U.S. is the only society which has no history (truly its own) from before the age of progress. English-speaking Canadians, such as myself, have despised and feared the Americans for the account of freedom in which their independence was expressed, and have resented that other traditions of the English-speaking world should have collapsed before the victory of that spirit; but we are still enfolded with the Americans in the deep sharing of having crossed the ocean and conquered the new land. All of us who came made some break in that coming. The break was not only the giving up of the old and the settled, but the entering into the majestic continent which could not be ours in the way that the old had been. It could not be ours in the old way because the making of it ours did not go back before the beginning of conscious memory. The roots of some communities in eastern North America go back far in continuous love for their place, but none of us can be called autochthonous, because in all there is some consciousness of making the land our own. It could not be ours also because the very intractability, immensity, and extremes of the new land required that its meeting with mastering Europeans be a battle of subjugation. And after that battle we had no long history of living with the land before the arrival of the new forms of conquest which came with industrialism.

That conquering relation to place has left its mark within us. When we go into the Rockies we may have the sense that gods are there. But if so, they cannot manifest themselves to us as ours. They are the gods of another race, and we cannot know them because of what we are, and what we did. There can be nothing immemorial for us except the environment as object. Even our cities have been encampments on the road to economic mastery.

It may be that all men are at their core the homeless beings. Be that as it may, Nietzsche has shown that homelessness is the particular mark of modern nihilism. But we were homeless long before the mobility of our mobilized technology and the mass nihilism which has been its accompaniment. If the will to mastery is essential to the modern, our wills were burnished in that battle with the land. We were made ready to be leaders of the civilization which was incubating in Europe.

The very use of the word "autochthonous" raises another way in which we are not Europeans. Living undivided from one's own earth: here is not only a form of living which has not been ours but which is named in a language the echoes of which are far from us. The remoteness of "chthonic" from us measures our separation from Europe. Greece lay behind Europeans as a first presence; it has not so lain for us. It was for them primal in the sense that in its perfected statements educated Europeans found the way that things are. The Greek writings bared a knowledge of the human and non-human things which could be grasped as firmness by the Europeans for the making of their own lives and cities. Most important, Plato and Aristotle presented contemplation as the height for man. Until Nietzsche, Socrates was known as the peak of Greekness.

To say this does not deny that there was for Europeans another primal — Christianity. Indeed the meeting of these two in men's lives, the manifold attempts to see them as one, to bring together contemplation and charity, the fact that they were seen by some to be antithetical and so either one or the other condemned, the way that each was interpreted and misinterpreted in terms of the other and each used against the other in the building of a civilization which was new and which was neither, these inter-relations formed the chief tension out of which Europe was shaped. It is still possible for some Europeans to live in one or the other as primal although they are part of a civilization which is so alien from both.

The degree to which the Greek was primal for Europeans can be seen in the fact that those theoretical men, from Machiavelli to Nietzsche, who delineated what modern Europe was to become when it was no longer explicitly Christian, made an increasing appeal to the Greeks as primal, while Christianity became for them either a boring, although necessary, convention, or an avowed enemy. Even as their delineation was founded on an increasingly radical criticism of Greek thought, they claimed to be rediscovering a more authentic account of what the ancients had meant than that held by their immediate predecessors; thus Machiavelli against the theologians, Rousseau against the English, Nietzsche against Rousseau and Hegel.[1] Even such a modern revolutionary as Saint-Just justified his use of terror by an appeal to classical sources. The ways of modern Europe have often been described as a species of secularized Christianity. However, the ambiguity remains: the

---

1 My understanding of this history is dependent on the writings of Mr. Leo Strauss. To express my enormous debt to that great thinker must not, however, hide the fact that I interpret differently the relation of Christianity to the modern philosophers.

formulations of modernity have often been made by men who claimed to be returning behind Christianity to the classics, and yet laid out a fundamental criticism of the classical accounts of science, art, politics, etc. And that criticism seems to have been influenced by the hidden depths of Biblical religion.

Members of the civilization which initiated modern technology often now express a fear of the Americanization of Europe, and state that fear in their identification of the U.S. with the pure will to technique. This may be an expression of their deeper fear that their own society in becoming sheerly modern has at last and perhaps finally lost touch with its primal and therefore perhaps with contemplation itself, and that thereby Europe, in its particularity, is no more.

For us the primal was much different. It was the meeting of the alien and yet conquerable land with English-speaking Protestants. Since the crossing of the ocean we have been Europeans who were not Europeans. But the Europeanness which remained for us was of a special kind because Calvinist Protestantism was itself a break in Europe — a turning away from the Greeks in the name of what was found in the Bible. We brought to the meeting with the land a particular non-Mediterranean Europeanness of the seventeenth century which was itself the beginning of something new.

To understand North America it is necessary to understand those Protestants and to understand particularly their connection to the new physical and moral science which were coming into being in Europe. Why was it that the new physical and moral sciences, although not initiated by Calvinists, found a particularly ready acceptance among them, especially among the Dutch and the English? Weber enucleated the central practical relation between capitalism and the Calvinists as the worldly asceticism of the latter. His exposition of the essentials of that relationship is true despite its mistakes in detail and his lack of theoretical depth. Marxist historians have taken up the subject and written clearly of the relation between the new capitalism and Puritanism, particularly as the two were linked together in the parliamentary party during the English Civil War.

Because they were concentrating on the practical relation between religion and society, neither Weber nor the Marxists were concerned with the deeper level of the matter, which is the connection between

Protestant theology and the new sciences. For example, more fundamental than the practical connections between capitalism, the parliamentary party, and Protestantism, lies the fact that the refugee Protestant theologians from the continent espoused so immediately the Baconian account of science and worked to make it influential in England. It is only possible to write here generally about the relation between Protestant theology and the new science. It sprang initially from one negative agreement: both the theologians and the scientists wished to free the minds of men from the formulations of mediaeval Aristotelianism, though for different reasons. Because of our present education, the criticism by the seventeenth century scientists of the traditional doctrines is well known. They criticized the mediaeval teleological doctrine with its substantial forms as preventing men from observing and understanding the world as it is. The criticism by the theologians is less well known and less easily understandable in an age such as ours. They attacked the mediaeval teleological doctrine as the foundation of what they called "natural" theology, and that theology was attacked because it led men away from fundamental reliance on Christian revelation. The teleological doctrine did this because it encouraged men to avoid the surd mystery of evil by claiming that final purpose could be argued from the world. Such mitigation led men away from the only true illumination of that mystery, the crucifixion apprehended in faith as the divine humiliation.[2]

But beyond this common negative attack on the mediaeval science, there was in the theology of the Calvinist Protestants a positive element which made it immensely open to the empiricism and utilitarianism

---

2 Luther laid down the whole of this with brilliant directness at the very beginning of the Reformation in some theses of 1518.

"Thesis 19. He is not worthy to be called a theologian who sees the invisible things of God as understood through the things that are made (Romans I. 20).

Thesis 20. But only he who understands the visible and further things of God through the sufferings and the Cross.

Thesis 21. The theologian of glory says that evil is good and good evil; the theologian of the Cross says that the thing is as it is:"

Luther, *Werke*, Weimar edition. Vol I, p. 354.

It is surely possible to see the relation of such a theological statement to later German philosophy.

in the English edition of the new sciences. Troeltsch has described that element and its consequent openness: "Calvinism, with its abolition of the absolute goodness and rationality of the Divine activity into mere separate will-acts, connected by no inner necessity and no metaphysical unity of substance, essentially tends to the emphasising of the individual and empirical, the renunciation of the conceptions of absolute causality and unity, the practically free and utilitarian individual judgement of all things. The influence of this spirit is quite unmistakably the most important cause of the empirical and positivist tendencies of the Anglo-Saxon spirit, which today find themselves in it as compatible with strong religious feeling, ethical discipline and keen intellectuality as they formerly did in Calvinism itself."[3] "Today" for Troeltsch was before 1914, so that "strong religious feeling, ethical discipline and keen intellectuality" must be taken as an account of the English-speaking bourgeois world before the adventures and catastrophes of the last half century, before the total collapse of Calvinism as an explicit social force. Indeed as Calvinism was more present in North America than in England as the dominant public religion, Troeltsch's words apply more forcibly to this continent than to the home of Puritanism.

This connection between the English-speaking Protestants and the new physical and moral sciences is played down by those who point to the worldliness of thinkers such as Hobbes and Locke, as compared to the stern account of salvation found among the Calvinists. Such a contrast is indeed obvious but misses the nature of the connection. It was not that the new philosophers were held by the truth of Christianity. Protestantism was merely a presence in the public world they inhabited which was more compatible with their espousings than Catholicism. Rather the connection was from the side of the Protestants who found something acceptable in the new ideas so that often they were the instruments for these ideas in the world, almost without knowing the results for their faith. At the least, Calvinist Christianity did not provide a public brake upon the dissemination of the new ideas as did Catholicism and even sometimes Anglicanism. For example, Locke, so important an influence on our North American destiny, may well be interpreted as contemptuous of Christian

---

3 E. Troeltsch, *Protestantism and Progress*, pp. 162–163.

revelation and even of theism itself. The comfortable self-preservation to which he thought men directed is hardly compatible with what any Christianity could assert our highest end to be. Nevertheless over the centuries it has been Protestants, both authentic and conventional, who have found his political and epistemological ideas so congenial. One of his great triumphs was surely that by the marvellous caution and indirectness of his rhetoric and by some changes of emphasis at the political level, he could make Hobbes' view of nature acceptable to a still pious bourgeoisie. Most of us do not see how our opinions are gradually changed from what we think we believe, under the influence of ideas elucidated by others incomparably deeper and more consistent than ourselves. "Worldly asceticism" was to become ever more worldly and less ascetic in the gradual dissolving of the central Protestant vision. The control of the passions in Protestantism became more and more concentrated on the sexual, and on others which might be conducive to sloth, while the passions of greed and mastery were emancipated from traditional Christian restraints. Weber was brilliantly right to place Franklin near the centre of his account of English-speaking Protestantism. Incomparably less philosophic than Locke, Franklin illustrates the influence back from Protestantism into the ideas of the new worldly modernity. He may have had contempt for revelation in his sensual utilitarianism, but the public virtues he advocates are unthinkable outside a Protestant ethos. The practical drive of his science beautifully illustrates what has been quoted from Troeltsch. It takes one quite outside the traditionally contemplative roots of European science, into the world of Edison and research grants. In 1968 Billy Graham at the Republican Convention could in full confidence use Franklin in his thanksgiving for what the Christian God had done for America.

The fact that such men have so often been the shock troops of the English-speaking world's mastery of human and non-human nature lay not simply in the absence of a doctrine of nature into which vacuum came the Hobbesian account of nature (so that when revelation was gone all that was left was that account) but also in the positive content of their extraordinary form of Christianity. The absence of natural theology and liturgical comforts left the lonely soul face to face with the transcendent (and therefore elusive) will of God. This will had to be sought and served not through our contemplations but directly through our practice. From the solitude and uncertainty of that position came the responsibility which could find no rest. That unappeasable responsibility gave an extraordinary sense of the self as radical freedom so paradoxically experienced within the predestinarian theological context. The external world was unimportant and indeterminate stuff (even when it was our own bodies) as compared with the soul's ambiguous encounter with the transcendent. What did the body matter; it was an instrument to be brought into submission so that it could serve this restless righteousness. Where the ordinary Catholic might restrain the body within a corporatively ordained tradition of a liturgy rhythmic in its changes between control and release, the Protestant had solitary responsibility all the time to impose the restraint. When one contemplates the conquest of nature by technology one must remember that that conquest had to include our own bodies. Calvinism provided the determined and organised men and women who could rule the mastered world. The punishment they inflicted on non-human nature, they had first inflicted on themselves.

Now when from that primal has come forth what is present before us; when the victory over the land leaves most of us in metropoloi where widely spread consumption vies with confusion and squalor; when the emancipation of greed turns out from its victories on this continent to feed imperially on the resources of the world; when those resources cushion an immense majority who think they are free in pluralism, but in fact live in a monistic vulgarity in which nobility and wisdom have been exchanged for a pale belief in progress, alternating with boredom and weariness of spirit; when the disciplined among us drive to an unlimited technological future, in which technical reason has become so universal that it has closed down on openness and awe, questioning and listening; when Protestant subjectivity remains authentic only where it is least appropriate, in the moodiness of our art and sexuality, and where public religion has become an unimportant litany of objectified self-righteousness necessary for the more anal of our managers; one must remember now the hope, the stringency and nobility of that primal encounter. The land was almost indomitable. The intense seasons of the continental heartland needed a people who whatever else were not flaccid. And these people not only forced commodities from the

land, but built public and private institutions of freedom and flexibility and endurance. Even when we fear General Motors or ridicule
our immersion in the means of mobility, we must not forget that the
gasoline engine was a needfilled fate for those who had to live in such
winters and across such distances. The Marxists who have described
the conquest of the continent as an example of capitalist rape miss
the substance of those events, as an incarnation of hope and equality
which the settlers had not found in Europe. Whatever the vulgarity
of mass industrialism, however empty our talk of democracy, it must
not be forgotten that in that primal there was the expectation of a
new independence in which each would be free for self-legislation, and
for communal legislation. Despite the exclusion of the African, despite
the struggles of the later immigrant groups, the faith and institutions
of that primal encounter were great enough to bring into themselves
countless alien traditions and make these loyal to that spirit. To know
that parents had to force the instincts of their children to the service
of pioneering control; to have seen the pained and unrelenting faces
of the women; to know, even in one's flesh and dreams, the results
of generations of the mechanising of the body; to see all around one
the excesses and follies now necessary to people who can win back
the body only through sexuality, must not be to forget what was
necessary and what was heroic in that conquest.

Now when Calvinism and the pioneering moment have both gone,
that primal still shapes us. It shapes us above all as the omnipresence
of that practicality which trusts in technology to create the rationalized kingdom of man. Other men, communists and national socialists,
have also seen that now is the moment when man is at last master of
the planet, but our origins have left us with a driving practical optimism which fitted us to welcome an unlimited modernity. We have
had a practical optimism which had discarded awe and was able to
hold back anguish and so produce those crisp rationalized managers,
who are the first necessity of the kingdom of man. Those uncontempla-
tive, and unflinching wills, without which technological society cannot
exist, were shaped from the crucible of pioneering Protestant liberalism.
And still among many, secularized Christianity maintains itself in the
rhetoric of good will and democratic possibilities and in the belief that
universal technical education can be kind, etcetera, etcetera. Santayana's

remark that there is a difference between Catholic and Protestant athe-
ism applies equally to liberalism; ours is filled with the remnential
echoes of Calvinism. Our belief in progress may not be as religiously
defined as the Marxist, but it has a freedom and flexibility about it which
puts nothing theoretical in the way of our drive towards it (or in other
words as the clever now say, it is the end of ideology). In short our very
primal allowed us to give open welcome to the core of the twentieth
century — the unlimited mastery of men by men.

It may be argued that other later arrivals from Europe have so placed
their stamp on North America as to have changed in essence what could
come from that primal. But obvious facts about the power of Catholicism
in our politics, or the influence of Jews in communications and intellectual
life, or the unexpected power for continuance shown by ethnic communities, mean only that recent traditions have coloured the central current of
the American Dream. The effectiveness of Catholics in politics remains
long after its origins in urban immigrant needs, but from the very beginning successful Catholic politicians have been particularly dutiful towards
institutions, customs, and rhetoric which had been made by others before
their arrival, and made from traditions utterly different from their own.
In so far as Catholic contemplation ever crossed the ocean, it has been
peripheral. Today when Catholics desiring to embrace the modern open
themselves directly to the public liberalism, it looks as if even the few
poor remnants of contemplation will die. For all the closeness to Jews to
the American Dream, it would be degrading to Judaism to say that it has
been able to express its riches in American culture when the chief public
contribution of Jews has been the packaged entertainment of Broadway
and Hollywood, the shallow coteries of intellectual New York. As for
pluralism, differences in the technological state are able to exist only in
private activities: how we eat; how we mate; how we practise ceremonies.
Some like pizza, some like steaks; some like girls, some like boys; some
like synagogues, some like the Mass. But we all do it in churches, motels,
restaurants indistinguishable from the Atlantic to the Pacific.

Even as the fissures in the system become apparent, leading its enemies to underestimate its ability to be the leader in modernity, our
primal spirit still partially survives to give our society its continuing
dynamism. The ruthlessness and banal callousness of what has been
done in Vietnam might lead one to see North American events as

solely self-interested nihilism of a greedy technological empire. But such an interpretation would not be sufficient to the reality. It must be remembered that the exigencies of imperialism have to be justified to the public (particularly to the second order managers) under the banner of freedom and a liberating mondernization. When they cannot there is widespread protest of a kind that never existed during the European depredations in the non-European world. The Vietnam War is disliked not only because it is obviously a tactical blunder; nor only because most of us are "last men" too comfortable to fight for the imperial power that buttresses that comfort; nor, simplistically, is it that television filters some of the ferocity to our living rooms; but also because the central dream still publicly holds, that North America stands for the future of hope, a people of good will bringing the liberation of progress to the world. The exigencies of violence necessary to our empire will increasingly make mockery of the rhetoric of that dream. The lineaments of our imperialism are less and less able to be dressed up in the language of liberal idealism to make them seem more than the affluence and power of the northern hemisphere. Nevertheless, as of now, the belief that America is the moral leader of the world through mondernization still sustains even the most banal and ruthless of our managers.

At home the ruling managers move "towards the year 2000." It might seem here that the practical primal has become no more than the unalloyed drive to technological mastery for its own sake. It is this interpretation which allows certain Europeans to consider us a wasteland with nothing seriously human amongst us but that self-propelling will to technology. But this interpretation underestimates the very effectiveness of North America in the world, in its forgetting that it is men who make that drive. What makes the drive to technology so strong is that it is carried on by men who still identify what they are doing with the liberation of mankind. Our ruling managers are able to do what they do just because among sufficient of them technology and liberalism support each other as identified. It is this identification which makes our drive to technology still more dynamic than the nihilistic will to will which is emptied of all conceptions of purpose. It may be (to use the indicative would be claiming to have grasped the very heart of what is) that this drive to practicality moves to become little more than a will to mastery governing the vacuous masses. But that is not yet how we understand our present. The identification in our practicality of masterful interference and the building of a human world still filters through the manifold structures of managerial and scientific Elite to be the governing faith of the society. All political arguments within the system, the squalls on the surface of the ocean (for example that about the rights of property in relation to the common good, between the freedom for some and the freedom for all) take place within the common framework that the highest good is North America moving forward in expansionist practicality. To think outside this faith is to make oneself a stranger to the public realm.

Indeed the technological society is not for most North Americans, at least at the level of consciousness, a "terra incognita" into which we must move with hesitation, moderation and in wonder, but a comprehended promised land which we have discovered by the use of calculating reason and which we can ever more completely inherit by the continued use of calculation. Man has at last come of age in the evolutionary process, has taken his fate into his own hands and is freeing himself for happiness against the old necessities of hunger and disease and overwork, and the consequent oppressions and repressions. The conditions of nature — that "otherness" — which so long enslaved us, when they appeared as a series of unknown forces, are now at last beginning to be understood in their workings so that they can serve our freedom. The era of our planetary domination dawns; and beyond that? That this is obviously good can be seen in the fact that we are able to do what we never could and prevent what we have never before prevented. Existence is easier, freer and more exciting. We have within our grasp the conquest of the problem of work-energy; the ability to keep ourselves functioning well through long spans of life and above all the overcoming of old prejudices and the discovery of new experiences, so that we will be able to run our societies with fewer oppressive authorities and repressive taboos.

To such comprehension the technological society is only in detail a terra incognita, as in its rushing change new problems arise which cannot always be predicted in advance. We therefore require the clearest minds to predict by understanding those which are on the horizon and to sort them out by calculation with courage. As we move "towards the year

2000" we need all the institutes of urban studies and of race relations, all the centres of economic development and psychological adjustment we can get. We will have to see how cities need not set affluence and squalor, private competence and public disorganization against each other; how all can reach a level of educational competence to inherit the hope; how the young can be shown purpose in the midst of enormous bureaucracies; how banality need not be incumbent on mass culture; how neuroses and psychoses, which are so immediately destructive when power is great, can be overcome by new understandings of psychology and sociology, etcetera, etcetera. Add to these the international problems of how underdeveloped countries can be brought to share in the new possibilities by accepting the conditions of mondernization, how the greed of already modern societies does not hold the others in slavery, how mass breeding with modern medicine does not overwhelm them and us before mondernization can be accomplished, above all how the new military techniques do not explode us all before we have reached an internationalism appropriate to the age of reason. But these are difficulties of detail, requiring our best calculation to avoid, but not vitiating intrinsically the vision of the technological society as a supreme step in our liberation. Behind them lie the comprehension of this great experiment in the minds of our dominant majority as self-evidently good, that for which man has struggled in evolution since his origins in pain and chance, ignorance and taboo.[4]

Indeed the loud differences in the public world — what in a simpler minded nineteenth century Europe could be described as the divisions between left and right — are carried on within this fundamental faith. The directors of General Motors and the followers of Professor Marcuse sail down the same river in different boats. This is not to say anything

---

4 As is true of all faiths, this dominating modern faith has many different expressions of itself. Some of these formulations put forward a rather low and superficial view of what it is to be human, for example those of Daniel Bell or Marion Levy in the U.S. or that of Edmund Leach in the U.K. These formulations must not lead to the Hermeneutical error of judging the truth of the faith from the crassness of a particular formulation. This would be as fair as judging the truth of Christianity from the writings of its most foolish theologians. The same modern faith has been expounded thoughtfully by many; by liberals, both positivist or existentialist, by Marxists, by Christians, and by Jews.

as jejeune as to deny the obvious fact that our technological society develops within a state capitalist framework and that that will have significant effect on what we are and what we will become, particularly in relation to other technological societies developed under other structures. But amid the conflict of public ideologies it is well to remember that all live within a common horizon. Those of the "right," who stand by the freedoms of the individual to hold property and for firmer enforcement of our present laws, seem to have hesitation about some of the consequences of modernity, but they do not doubt the central fact of the North American dream — progress through technological advance. It may be indeed that, like most of us, the "right" want it both ways. They want to maintain certain moral customs, freedoms of property, and even racial rights which are not in fact compatible with advancing technological civilization. Be that as it may, the North American "right" believes firmly in technical advance. Indeed its claim is that in the past the mixture of individualism and public order it has espoused has been responsible for the triumphs of technique in our society.[5]

Equally those of the "left" who have condemned our social arrangements and worked most actively to change them have based their condemnation in both the 1930s and 1960s on some species of Marxism. This is to appeal to the redemptive possibilities of technology and to deny contemplation in the name of changing the world. Indeed domestic Marxists have been able as a minority to concentrate on the libertarian and utopian expectations in their doctrines because unlike the Marxists of the East they could leave the requirements of public order to others. But however libertarian the notions of the new left, they are always thought within the control of nature achieved by modern techniques. The liberation of human beings assumes the ease of an environment where nature has already been conquered. For example, at the libertarian height of Professor Marcuse's writings (*Eros and Civilization*),

---

5 I use the term "right" because I have written elsewhere of the impossibility of political conservatism in an era committed to rapid technological advance. See *Lament for a Nation*, pp. 66–67. The absurdity of the journalistic use of the word conservative was seen in the reporting of the recent invasion of Czechoslovakia when the term conservative was widely applied to the pro-Russian Czech communist leaders.

he maintains that men having achieved freedom against a constraining nature can now live in the liberation of a polymorphous sexuality. The orgiastic gnosticism there preached always assumes that the possibilities of liberation depend on the maintenance of our high degrees of conquest. Having first conquered nature we can now enjoy her. His later *One Dimensional Man* is sadder in its expectations from our present situation, but technology is still simplistically described and blessed, as long as it is mixed with the pursuit of art, kind sexuality, and a dash of Whiteheadian metaphysics.

Even the root and branch condemnation of the system by some of the politicized young assumes the opportunities for widespread instant satisfaction which are only possible in terms of the modern achievements. They want both high standards of spontaneous democracy and the egalitarian benefits accruing from technique. But have not the very forms of the bureaucratic institutions been developed as necessary for producing those benefits? Can the benefits exist without the stifling institutions? Can such institutions exist as participatory democracies? To say yes to these questions with any degree of awareness requires the recognition of the fact that the admired spontaneity of freedom is made feasible by the conquering of the spontaneity of nature. In this sense their rejection of their society is not root and branch. They share, with those who appear to them as enemies, the deeper assumptions which have made the technological society.

Indeed the fact that progress in techniques is the horizon for us is seen even in the humane stance of those who seek some overreaching vision of human good in terms of which the use of particular techniques might be decided. Who would deny that there are many North Americans who accept the obvious benefits of modern technique but who also desire to maintain firm social judgement about each particular method in the light of some decent vision of human good? Such judgements are widely attempted in obvious cases, such as military techniques, where most men still ask whether certain employments can ever serve good. (This is even so in a continent whose government is the only one so far to have used nuclear weapons in warfare.) At a less obvious level, there are still many who ask questions about particular techniques of government planning and their potency for tyranny. Beyond this again there are a smaller number who raise questions about new biochemical methods and their relation to the propagation of the race. As the possible harm from any new technique is less evident, the number of questioners get fewer. This position is the obvious one by which a multitude of sensible and responsible people try to come to terms with immediate exigencies. Nevertheless the grave difficulty of thinking a position in which technique is beheld within a horizon greater than itself, stems from the very nature of our primal, and must be recognized.

That difficulty is present for us because of the following fact: when we seek to elucidate the standards of human good (or in contemporary language "the values") by which particular techniques can be judged, we do so within modern ways of thought and belief. But from the very beginnings of modern thought the new natural science and the new moral science developed together in mutual interdependence so that the fundamental assumptions of each were formulated in the light of the other. Modern thought is in that sense a unified fate for us. The belief in the mastering knowledge of human and non-human beings arose together with the very way we conceive our humanity as an Archimedean freedom outside nature, so that we can creatively will to shape the world to our values. The decent bureaucrats, the concerned thinkers, and the thoughtful citizens as much conceive their task as creatively willing to shape the world to their values as do the corporate despots, the motivations experts and the manipulative politicians. The moral discourse of "values" and "freedom" is not independent of the will to technology, but a language fashioned in the same forge together with the will to technology. To try to think them separately is to move more deeply into their common origin.

Moreover, when we use this language of "freedom" and "values" to ask seriously what substantive "values" our freedom should create, it is clear that such values cannot be discovered in "nature" because in the light of modern science nature is objectively conceived as indifferent to value. (Every sophomore who studies philosophy in the English-speaking world is able to disprove "the naturalistic fallacy," namely, that statements about what ought to be cannot be inferred solely from statements about what is.) Where then does our freedom to create values find its content? When that belief in freedom expresses itself seriously

21

(that is politically and not simply as a doctrine of individual fulfilment) the content of man's freedom becomes the actualising of freedom for all men. The purpose of action becomes the building of the universal and homogeneous state — the society in which all men are free and equal and increasingly able to realize their concrete individuality. Indeed this is the governing goal of ethical striving, as much in the modernizing east as in the west. Despite the continuing power in North America of the right of individuals to highly comfortable and dominating self-preservation through the control of property, and in the communist bloc the continuing exaltation of the general will against all individual and national rights, the rival empires agree in their public testimonies as to what is the goal of human striving.

Such a goal of moral striving is (it must be repeated) inextricably bound up with the pursuit of those sciences which issue in the mastery of human and non-human nature. The drive to the overcoming of chance which has been the motive force behind the developers of modern technique did not come to be accidentally, as a clever way of dealing with the external world, but as one part of a way of thought about the whole and what is worth doing in it. At the same time the goal of freedom was formulated within the light of this potential overcoming of chance. Today this unity between the overcoming and the goal is increasingly actualized in the situations of the contemporary world. As we push towards the goal we envisage, our need of technology for its realization becomes ever more pressing. If all men are to become free and equal within the enormous institutions necessary to technology, then the overcoming of chance must be more and more rigorously pursued and applied — particularly that overcoming of chance among human beings which we expect through the development of the modern social sciences.

The difficulty then of those who seek substantive values by which to judge particular techniques is that they must generally think of such values within the massive assumptions of modern thought. Indeed even to think "values" at all is to be within such assumptions. But the goal of modern moral striving — the building of free and equal human beings — leads inevitably back to a trust in the expansion of that very technology we are attempting to judge. The unfolding of modern society has not only required the criticism of all older standards of human excellence, but has also at its heart that trust in the overcoming of chance which leads us back to judge every human situation as being solvable in terms of technology. As moderns we have no standards by which to judge particular techniques, except standards welling up with our faith in technical expansion. To describe this situation as a difficulty implies that it is no inevitable historicist predicament. It is to say that its overcoming could only be achieved by living in the full light of its presence.

Indeed the situation of liberalism in which it is increasingly difficult for our freedom to have any content by which to judge techniques except in their own terms is present in all advanced industrial countries. But it is particularly pressing for us because our tradition of liberalism was moulded from practicality. Because the encounter of the land with the Protestants was the primal for us, we never inherited much that was at the heart of western Europe. This is not to express the foolish position that we are a species of Europeans-minus. It is clear that in our existing here we have become something which is more than European — something which by their lack of it Europeans find difficult to understand. Be that as it may, it is also clear that the very nature of the primal for us meant that we did not bring with us from Europe the tradition of contemplation. To say contemplation "tout court" is to speak as if we lacked some activity which the Ford Foundation could make good by proper grants to the proper organizations. To say philosophy rather than contemplation might be to identify what is absent for us with an academic study which is pursued here under that name. Nevertheless, it may perhaps be said negatively that what has been absent for us is the affirmation of a possible apprehension of the world beyond that as a field of objects considered as pragmata — an apprehension present not only in its height as "theory" but as the undergirding of our loves and friendships, of our arts and reverences, and indeed as the setting for our dealing with the objects of the human and non-human world. Perhaps we are lacking the recognition that our response to the whole should not most deeply be that of doing, nor even that of terror and anguish, but that of wondering or marvelling at what is, being amazed or astonished by it, or perhaps best, in a discarded English usage, admiring it; and that such a stance, as beyond all bargains and conveniences, is the only source from which purposes may be manifest to us for our necessary calculating.

To repeat, western Europe had inherited that contemplation in its use of it theologically, that is, under the magistery of revelation. Within that revelation charity was the height and therefore contemplation was finally a means to that obedient giving oneself away. Nevertheless it was necessary for some to think revelation and the attempt to do so led theologians continually back to the most comprehensive thinkers that the west had known. Augustine spoke of "spoiling the Egyptians" but in that use of philosophy to expound revelation, the spoilers were often touched by that which they would use as something they could not use. In that continual tasting of the Greeks, some men were led back to thought not determined by revelation, and therefore to a vision of contemplation not subservient to charity, but understood as itself the highest. As has been said earlier, the Calvinists claimed to be freeing theology from all but its Biblical roots and cut themselves off from pure contemplation more than did any other form of European theology — Catholic or Jewish, Lutheran or even Anglican. For the Calvinist, theology was a prophetic and legal expounding of a positively conceived revelation, the purpose of which was to make its practical appeal to men. Thus being in our origins this form of Protestant, thrown into the exigencies of the new continent, we did not partake of the tradition of European contemplation. And as we moved from that Calvinism to modernity, what was there in the influence of liberalism which could have made us more open to that contemplation? Indeed for lack of contemplation, American intellectual patriots have had to make the most of Emerson and Adams, James and Pierce.

I know how distant from North Americans is the stance of contemplation, because I know the pervasiveness of the pragmatic liberalism in which I was educated and the accidents of existence which dragged me out from it. To write so may seem some kind of boasting. But the scavenging mongrel in the famine claims no merit in scenting food. Perhaps for later generations of North Americans it is now easier to turn and partake in deeper traditions than they find publicly around them. The fruits of our own dominant tradition have so obviously the taste of rot in their luxuriance. It may be easier for some of the young to become sane, just because the society is madder. But for myself it has taken the battering of a lifetime of madness to begin to grasp even dimly that which has been inevitably lost in being North American. Even to have touched Greekness (that is to have known it not simply as antiquarianism) required that I should first have touched something in Europe which stayed alive there from before the age of progress through all its acceptance of that age. By touching Europe I do not mean as a fascinating museum or a place of diversion, but to have felt the remnants of a Christianity which was more than simply the legitimizing of progress and which still held in itself the fruits of contemplation. By that touching I do not mean the last pickings of authentic theology left after the storms of modern thought (though that too) but things more deeply in the stuff of everyday living which remain long after they can no longer be thought: public and private virtues having their point beyond what can in any sense be called socially useful; commitments to love and to friendship which lie rooted in a realm outside the calculable; a partaking in the beautiful not seen as the product of human creativity; amusements and ecstasies not seen as the enemies of reason. This is not to say that such things did not or do not exist in North America (perhaps they cannot disappear among human beings) but their existence had been dimmed and even silenced by the fact that the public ideology of pragmatic liberalism could not sustain them in its vision. The remnants of that which lay beyond bargaining and left one without an alternative still could be touched even amidst the degeneracy of Europe's ruin. They generally existed from out of a surviving Christianity or Judaism (neither necessarily explicit) which pointed to a realm in which they were sustained. I remember the surprise — the distance and the attraction — of letting near one at all seriously a vision of life so absent in day to day North America. I remember how such a vision inevitably jeopardised one's hold on North America: how it made one an impotent stranger in the practical realm of one's own society. But the remnants of such a Europe were only one remove from what was one's own. It was the seedbed out of which the attenuated Christianity of our secularized Calvinism had come. To touch the vestiges of this fuller Christianity was a possible step in passing to something which was outside the limits of one's own.

Indeed until recently the very absence of a contemplative tradition spared us the full weight of that public nihilism which in Europe flowered with industrial society. The elimination of the idea of final purpose from the scientific study of the human and non-human things not only led to the progress of science and the improvement of conditions but also had consequences on the public understanding of what it was to live. But this consequence was not so immediately evident in our practical culture as it was to Europeans. We took our science pragmatically, as if its effect on us could be limited to the external. Thus it was possible for us to move deeply into the technological society, while maintaining our optimism and innocence.

In the public realm, this optimism and innocence delayed the appearance amongst us of many of those disorders which in Europe were concurrent with that nihilism. It is well to remember that large sections of our population resisted the call to imperialism by the economic and political powers of the eastern seaboard, even when they welcomed the technological expansion which made it inevitable. Europeans (particularly the English) would do well to remember, now that they live in the full noon of that imperialism, how hard they worked to drag North American democracy to wider imperial pursuits. Until recently there have not appeared amongst us those public atheisms of the left and of the right which were central to the domestic violence of Europe in this century. The propertied classes of the right have remained uneducated until recently and so kept longer within the respectable religion of their tradition than did their counterparts in Europe. Liberals have ridiculed as hypocrisy the continuing religion among the propertied and even among the bureaucratic. When such traditions have gone those ridiculers may miss the restraints among their rulers that were part of such traditions. For can there be any doubt that the bureaucratic "right" must be more powerful in advanced societies than the left? For the last hundred years our optimism has been reaffirmed by generations of new immigrants who, whatever their trials, found in the possibilities of the new land the opportunity of affluence and freedom on its practical terms. This continuous entry of new families and new peoples busy fighting to partake in the North American dream perpetuated the vitality of the modern.

Even as the language of Europe's "agony" began to penetrate our institutions of the intellect, we were able to use that language as if it could be a servant of our optimistic practical purposes. To repeat, what would North American rhetoric be without the word "values"? But even those who use the word seriously within theoretical work seem not to remember that the word was brought into the centre of western discourse by Nietzsche and into the discourse of social science through Nietzsche's profound influence upon Weber. For Nietzsche the fundamental experience for man was apprehending what is as chaos; values were what we creatively willed in the face of that chaos by overcoming the impotence of the will which arises from the recognition of the consequences of historicism. Nietzsche's politics (and he affirmed that the heart of any philosophy can be seen in its political recommendations) stated that democracy and socialism were the last debasements brought into the world by Christianity as it became secularized. The universal and homogeneous state would be made up of "last men" from whom nobleness and greatness would have departed. Because of our firm practicality, North American social scientists have been able to use the language of values, fill it with the substantive morality of liberalism and thereby avoid facing what is assumed in the most coherent unfolding of this language. The writings of Lasswell and Parsons were hymns to that innocent achievement. It has been wonderful to behold legions of social scientists wising up others about the subjectiveness of their values while they themselves earnestly preached the virtues of industrial democracy, egalitarianism, and decent progressive education; espousing, in other words, that liberalism which sees the universal and homogeneous state as the highest goal of political striving. They took their obligations to the indigenous traditions more seriously than those to the theoretical consequences of their sciences.

Such a position could not last. The languages of historicism and values which were brought to North America to be the servants of the most advanced liberalism and pluralism, now turn their corrosive power on our only indigenous roots—the substance of that practical liberalism itself. The corrosions of nihilism occur in all parts of the community. Moreover, because our roots have been solely practical, this nihilism shares in that shallowness. The old individualism of capitalism, the frontier and Protestantism, becomes the demanded right to one's idiosyncratic wants taken as outside any obligation to the community which provides them. Buoyed by the restless needs of affluence, our art becomes hectic in its experiments with style and violence. Even the surest accounts of our technomania — the sperm-filled visions of Burroughs — are themselves spoken from the shallowness they would describe. Madness itself can only be deep when it comes forth from a society which holds its opposite. Nihilism, which has no tradition of contemplation to beat against, cannot be the occasion for the amazed reappearance of the "What for? Whither? and What then?". The tragedy for the young is that when they are forced by its excesses to leave the practical tradition, what other depth is present to them in which they can find substance? The enormous reliance on and expectation from indigenous music is a sign of the craving for substance, and of how thin is the earth where we would find it. When the chthonic has been driven back into itself by the conquests of our environment, it can only manifest itself beautifully in sexuality, although at the same time casting too great a weight upon that isolated sexuality.

For those who stay within the central stream of our society and are therefore dominant in its institutions, the effect of nihilism is the narrowing to an unmitigated reliance on technique. Nietzsche's equivocation about the relation between the highest will to power and the will to technology has never been part of the English speaking tradition. With us the identity was securely thought from the very beginning of our modernity. Therefore as our liberal horizons fade in the winter of nihilism, and as the dominating amongst us see themselves within no horizon except their own creating of the world, the pure will to technology (whether personal or public) more and more gives sole content to that creating. In the official intellectual community this process has been called "the end of ideology." What that phrase flatteringly covers is the closing down of willing to all content except the desire to make the future by mastery, and the closing down of all thinking which transcends calculation. Within the practical liberalism of our past, techniques could be set within some context other than themselves — even if that context was shallow. We now move towards the position where technological progress becomes itself the sole context within which all that is other to it must attempt to be present.

We live then in the most realized technological society which has yet been; one which is, moreover, the chief imperial centre from which technique is spread around the world. It might seem then that because we are destined so to be, we might also be the people best able to comprehend what it is to be so. Because we are first and most fully there, the need might seem to press upon us to try to know where we are in this new found land which is so obviously a "terra incognita." Yet the very substance of our existing which has made us the leaders in technique, stands as a barrier to any thinking which might be able to comprehend technique from beyond its own dynamism.

Religion and the State

INTRODUCTION

THIS ESSAY IS INCLUDED *to illustrate two points: (1) the futility of conservatism as a theoretical standpoint in our era, (2) the degree to which hopes and prejudices, desires and vices cloud the mind when one tries to think about important matters. Luckily the analogy between the fruits of one's loins and the fruits of one's pen cannot be carried too far. If a child grew up to be a fatuous adult, it would be a matter of sorrow; the survival of an obtuse writing is only a matter of amusement.*

The folly of this writing is that it did not grasp what the technological society really is. Therefore the general principles of political philosophy asserted in it have no possible application in the society to which it is addressed. This is fatal for an article on a practical matter. It is like those reforming women who go on delegations to tell our rulers that the foreign policy of our empire should be based on the principles of the United Nations. Their virtuous proposals have no point because they are abstracted from the situation as it is. Most of the general propositions appear to me true — (for example, that reverence rather than freedom is the matrix of human nobility). But to think there is a point in saying any of these things as if they could have public relevance in the English-speaking world of the twentieth century is absurd.

What caused this absurdity was the following: faced at an early age by the barrenness of the all-pervading liberalism, I had spent much of my life looking for a more adequate stance. In doing so I had touched wonderful truths from our origins in Athens and Jerusalem. But to write as if these could be "conservatively" appropriated or publically sustained in our present society showed that I understood this society so little as to have no business writing about it. To partake even dimly in the riches of Athens or Jerusalem should be to know that one is outside the public realm of the age of progress. The relation between society and the call to philosophy must never be seen as accidental, because only in and through what is now present for us can the most important questions make their appearance. Since our society is technological those questions cannot appear if clouded by any "conservative" hope. In saying this I do not deny that conservatism is a noble practical stance.

Here point (1) and point (2) meet. Hopes and prejudices, desires and vices come in to cloud thinking about what is. If one is raised in the North American dream one so wants one's society and its institutions to have potentialities for nobility. For example, I hoped for years that our ecclesiastical organizations (being the guardians of the beauty of the gospel) might continue to be able to permeate this society with something nobler than the barrenness of technical dynamism. I hoped for this when every piece of evidence before me was saying that it was not true. I could not face the fact that we were living at the end of western Christianity. I could not believe that the only interpretation of Christianity that technological liberalism would allow to survive publically would be that part of it (e.g., the thought of Teilhard) which played the role of flatterer to modernity. Beyond such foolish hope lay the vice of ambition. One wants one's thoughts to be influential. Thinking in any era requires courage to sustain it. But courage always tends to fall over into ambition and as such corrupts the very thinking that courage must sustain. To want one's thoughts about the practical to be influential can lead to this corrupting ambition.

There would be little point in republishing this essay simply as an illustration of my own changes in thought or my particular vices, but something like this happens in all open thinking. In our age official thought exalts shallow positivist competence and the belief in mathematicising as the objective way. Indeed some of our potentially clearest thinkers turn to this triumph of algebra just because they want to be freed from the uncertainties of prejudice and desire which threaten open thought. By thinking only about what is mathematicisable they can abstract from themselves and their own ambiguities, into the safe light of the quantifiable object.

The opposite to such positivist competence is openness to all that is. If such openness is to confront the shallow public competence, it must include an openness to the distortions of our social prejudices and our tortured instincts. This does not imply the current belief that either sociology or psychology is the magisterial science to philosophy or theology — quite the contrary. But if philosophy is to transcend sociology or psychology, it must hold within itself any proper therapy which comes from such sciences. To move "ex umbris et imaginibus in veritatem" involves continually bringing to consciousness all the distortions which are bound to be present from one's individual and social history.

THE PRESENT CONTROVERSY ABOUT the proper place of religious education in the public schools of Ontario has been argued with passion by the various parties to the debate. The presence of this passion is hardly surprising since the proponents on both sides have believed that they are defending important principles. On the one hand, those who advocate that there should be no formal religious instruction in the schools believe that they stand on the principles of liberalism — the division of state and church, the freedom of religious minorities, the right of the individual to work out his own opinions. On the other hand, those who advocate the continuation of religious instruction in our schools believe that they are defending the place of Christianity in our national life at a time of spiritual chaos and that Christianity is the pearl of great price upon which all that is valuable in western civilization depends. In such a situation, the passion of the controversy is hardly unexpected.

The intensity of the debate has not been matched, however, by an attempt on the part of the protagonists of either side to state systematically the principles which should govern the relation of religion to the state and an attempt to apply those principles to our present situation. The result is that the controversy has taken the form of a struggle for political power rather than a debate between members of a common society. It has, therefore, been marked by an absence of good feeling. This is perhaps to be expected in an age which more and more glorifies "decisiveness" in politics at the expense of "thoughtfulness." But the price of decisiveness is often bitterness and surely nowhere is bitterness more to be eschewed than in questions of the social place of religion. The purpose of the present article is to raise doubts about certain of the arguments employed by both sides in the present controversy. Such an approach may be labelled as negative and unconstructive. It is surely the case, however, that in matters as fundamental as the role of religion in society, it is better to raise objections to superficial solutions than to advocate passionate solutions based on confused principle. In any case, it is the function of the practising politician to seek compromise solutions in different cases; it is the function of the philosopher to argue general principles.

As a preliminary, it is necessary to state how the word "religion" will be used in this article. The central controversy about the use of the word has been whether it should be confined to those systems of belief which include reference to a "higher" divine power or whether it should be extended to include those beliefs which exclude any reference to such a power. I intend to use it in the latter and broader sense. The origin of the word is, of course, shrouded in uncertainty, but the most likely account is that it arises from the Latin "to bind together." It is in this sense that I intend to use it. That is, as that system of belief (whether true or false) which binds together the life of individuals and gives to those lives whatever consistency of purpose they may have. Such use implies that I would describe liberal humanists or Marxists as religious people; indeed that I would say that all persons (in so far as they are rational beings) are religious. It is impossible outside a treatise on the philosophy of religion to justify this broader use of the term as against the more limited one. I can, however, raise one difficulty about the narrower definition which leads to suspicion of its use. Are we not to call Buddhism, or Marxism, religions? Yet neither of these in their purest form make any reference to a "higher" divine power.

In terms of this usage, I will first raise a difficulty in the argument employed by those who advocate the elimination of religious instruction in the schools and then a difficulty in the argument of those who advocate its continuance. The most generally used argument of those who are against the present system of religious education is that the state should eschew any involvement in the religious opinion of its members because the modern democratic state is committed to pluralism of opinion. It is the function of the family in co-operation with various religious institutions (including private denominational schools, if necessary) to teach their own youngsters their particular religious tradition. The state should only interfere with religious belief when the opinions of some of its members threaten outright the safety of others or threaten the proper authority of the state itself. This removal of the state from the positive sphere of religious instruction is necessary to freedom of opinion and to the separation of church and state — mutually interdependent principles which are of the very substance of democratic government.[1]

My criticism of this argument would be that it fails to come to grips with the fact that religion has a public as well as a personal role. If religion were simply a personal concern, the above argument would be irrefutable. But religion has more than a personal role because the state has an indubitable, if limited, interest in what its members believe. That such an interest exists can be demonstrated in the following way. Constitutional government has an interest in maintaining public order with a minimum use of coercive power.[2] Indeed constitutional government can hardly continue to exist if society has reached the point where the state can only maintain its proper authority (and through it, public order) by the widespread exercise of police power rather than by the free consent of the majority of its citizens. But the free acceptance of a certain minimum of public order on the part of the citizens depends on what those citizens believe to be true about their lives. For example, if citizens come to believe the religion that violence is an end in itself (as was the case with an influential number of Germans after 1890) they cannot be

---

1 I do not intend to discuss here the historical problem of whether Canada is committed to the same interpretation of the principle of the separation of church and state as is the American republic. This has become now largely an antiquated question as our indigenous traditions have been worn away by the increasing waves of American imperialism.

2 I have limited my argument to constitutional government because I take for granted that it is not necessary at this time in Canada to argue that constitutional government is superior to tyrannical government. Also, even though tyrannies depend to some extent on the beliefs of the members of the state, the dependence of the constitutional state is far greater because of its necessary hesitation to use coercion except as a last resort.

expected to remain loyal to constitutional government. Or again (and more apposite to our present situation in Canada) if citizens come to believe that their immediate desires must be satisfied even if it involves breaking the law and will therefore break the law unless they are afraid of being caught, this inevitably brings an increase in police power, to an extent which cannot but be inimical to constitutional government. The proposition may then be asserted that the state has an interest in the beliefs of its members in so far as those beliefs have bearing on the maintenance (and indeed perhaps even the improvement) of public order and on the authority of the state necessary to its survival.

The constitutional state then has an interest in limiting pluralism of belief: that limitation being what is necessary to the continuance of constitutional government. The question arises: what beliefs are necessary to that minimum public morality without which constitutional government is not possible? Does this minimum morality rest upon belief in a higher divine power? It is serious difference of opinion about this last question which underlies our present dispute about public education and religion. For it is surely clear that if constitutional government can hope to exist without such belief among the citizens, then the state has no interest nor business in teaching the basis of piety in its schools. If, on the other hand, constitutional government requires such public belief from a large percentage of its members, then it is the duty and interest of the state to use its schools to support the continuance of religious tradition.

Indeed the present controversy is not concerned with whether religion should be taught in the schools, but rather with what should be the content of the religion that is so taught. It is perfectly clear that in all North American state schools religion is already taught in the form of what may best be called "the religion of democracy." That the teaching about the virtues of democracy is religion and not political philosophy is clearly seen from the fact that the young people are expected to accept this on faith and cannot possibly at their age be able to prove the superiority of democracy to other forms of government (if indeed this can be done). The fact that those liberals who most object to any teaching about the deity are generally most insistent that the virtues of democracy be taught, should make us aware that what is at issue is not religion in general, but the content of the religion to be taught. Indeed I think that such liberal people are probably right that instruction in the faith of democracy should be carried on. In all times and places constitutional government is not necessarily democratic government. But in North America our very heritage of legal government is mutually interdependent with our heritage of democracy. Therefore it seems necessary that faith in that heritage should be widely taught. But let us be clear that not all parents take easily to such teaching. For instance, I, as a parent and a Christian, keep careful watch on the inculcation of this democratic faith among my children in case they should confuse their loyalty to a particular ordering of this passing world with the absolute loyalty which they owe to that which is beyond the world. Also I watch this inculcation in the schools, just as I watch their Sunday schools, to see that the myths proper to the young are not so crude as to be inimical to their growing faculty of rational judgement. Nevertheless I would start no campaign against the propagation of the democratic myth in our schools because I recognize that it may be necessary to the public good in the mass age.

The foregoing illustration was used to make clear that the question presently at issue is not whether religion should be taught in the state schools but rather whether what is taught should include any reference to the deity. When the question is put in this form, it is clearly a difficult one. Does public morality rest on the widespread practice of piety? This question is difficult because it is so broad and because it is one which has divided the modern world at the subtlest level of principle. To put the matter historically: from the dawn of western civilization until the nineteenth century the consensus was that piety was necessary to the public good. Any questioning of this assumption was the work of a small minority. To state the obvious, this opinion was strenuously held in the pre-Christian as well as the Christian era. In the last centuries this proposition has come under criticism by leading western minds, and since the nineteenth century this criticism has filtered out from the élite to become part of the consciousness of large numbers of citizens. Indeed one fundamental mark of the worldly faith which is so prevalent in our age is the opinion that the public good requires the inculcation of socially useful passions, and that the inculcation of such passions does not in any sense require the encouragement of the practice of piety.

This profound difference about what constitutes the basis of a constitutional public order underlies the present issue in Ontario far more deeply than differences about the proper relation of church and state. It is, however, quite impossible in the space of an article to describe fairly the arguments on either side concerning the matter. That the correct answer is not immediately evident is surely vouchsafed in the fact that political philosophers of the order of Plato and Spinoza can disagree on the question. I am not going to be as presumptuous as to reduce the subtle arguments on either side to a few platitudinous sentences and then quickly express my agreement with the tradition of piety. What is important in the present controversy is the recognition that disagreement about this matter is the basic cause of difference.

It is however necessary to add the qualification which the older tradition appended to the proposition that the good ordering of society required the practice of religion. The older political philosophers such as Plato, the Jewish Aristotelians, or Aquinas, asserted that unassisted reason was able to perceive that without religious beliefs and actions no society whatever can last; but they also asserted that unassisted reason is unable to determine what the religious beliefs and practices should specifically be.[3] That is, without divine revelation men could not know what should be the public religion — let alone the true religion. It was indeed only the certainty on the part of most members of society and more particularly of most political leaders that divine revelation had been granted to the Christian Church that resulted in the fact that that religion was the public religion through most of the European era.

This qualification leads to a distinction which seems to me essential for understanding society; namely, that between the true religion and the public religion. This distinction does not imply that all men think there is such a thing as the true religion or that those who think that there is such a thing are agreed as to its nature. Neither does it imply that all men think that there should be a public religion. It simply implies that some men have believed that they could know that there is such

---

3 It is impossible in short space to discuss the concept "unassisted reason."
I will therefore accept it to mean what it has meant generally in the
western tradition up to about 1700.

a thing as the true religion and that most societies (if not all) have not been entirely pluralist. This qualification also implies that when one particular religion is the public religion of a society, this is possible because a large percentage of the dominant classes in that society think it is the true religion. The right of a religion to public status is therefore a right of tradition and based on particular circumstances. It is also true that when particular religions have claimed and received public status, the rights they have afforded members of other religions in that public domain have depended on the intrinsic character of the public religion.

Those people who argue against the present arrangements seem to me sometimes confused as to whether they are appealing to the general principle that the good society does not need any widespread belief in a "higher" divine power, or whether they are appealing to the qualification that reason cannot tell us which specific religion should be publicly encouraged. If they appeal to the first, then surely they must expect to meet the full weight of conservative opinion that piety is necessary to the proper ordering of society, and they have no right to accuse (as they have done) such opinion of being simply bigoted and superstitious. If, on the other hand, they appeal to the qualification, then they may expect much more wide support for their attack on the present arrangements. Their confusion of the two different arguments has added to the confusion of the present situation. For example, it is surely the case that religious conservatives can be much more sympathetic to the attack on the present arrangements launched by the Jewish community than to that launched by the liberal secularists.

It is necessary to insist on this qualification because it is related also to the weakness in the argument used by some of my fellow Christians in defence of the present arrangement. This argument generally takes somewhat the following shape: (a) the good ordering of society requires religion, (b) the majority of Canadians are Christians and that therefore (c) the religion taught in the state schools should be Christian. As I have said, I do not intend to argue the truth of proposition (a), although there are very good grounds in political philosophy for asserting it and I myself believe it to be true. It is rather proposition (b) that needs very careful scrutiny.

In what sense can the appeal be made that the majority and dominant section of the community are Christians? In such a difficult question the appeal must be to impressions. There is the Roman Catholic community and also the strong influence of the Protestant sects; both influences seem to be growing in the industrial communities (a fact occasioning surprise for progressive liberals). There is also the continuing tradition of rural Protestantism, supplemented by the rooted people in the towns, often professional, who follow what they have inherited. There is also the growth of suburban Christianity since 1945, the meaning and consequences of which are difficult to assess. On the other side of the picture there are the large number of industrial workers whose situation has left them indifferent to their past and who are indeed indifferent to nearly all the established institutions of society. More important is the fact that an increasing percentage of the educated classes have since 1914 been intellectually cut off from Christianity. This is not only because of the fact that the universities have become increasingly disseminators of secularist ideas, but because the very *Weltanschauung* of the western world (in which the educated partake more than other parts of society) has been overwhelmingly secularist. Indeed the two factors go closely together. How could the universities be anything but disseminators of secularism, when one of their chief functions is the dissemination of the most recent ideas, techniques, and discoveries and these (particularly in the social sciences) have been overwhelmingly secularist? The result of this has been that among the educated classes of our society there is ever increasing weakening of their participation in a real Christianity.

This has not meant, of course, that many of the educated classes do not go to church. Of course they do, because they see with clarity the need of a moral and religious tradition in their society and in particular for their children. Two of the most successful and representative political leaders of the bureaucratic bourgeoisie in North America have well illustrated this position. Former President Eisenhower stated clearly during his term that he wanted American leaders to be religious, but he didn't much care about the content of that religion. Premier Leslie Frost entered decisively in favour of religious education in 1961 when he made clear that religion in his opinion was necessary for the balanced ethical life in the mass scientific society. These two men have been politically representative of the responsible business-government élite who direct the state capitalism which presently determines the tone of our society and which is likely to continue to govern us in the years ahead.

The majority of this class see clearly the need of religion in our society. As the religious side of them is the traditionalist side, they take for granted that the public religion will have something to do with Christianity. But they are not concerned with the clear definitions of theology about dogma and ritual which must continually be made by the Christian Church, and which alone can guarantee that the religion called Christian will in fact be Christian. That is, they are quite uncertain about the central question of Christianity — whether revelation of a decisive nature has once and for all been given. When they think of theology at all they think of it as the vested professional interest of the clergy which the laymen have to put up with as best they may. This ambiguous relation to the Christian religion is well illustrated by what the representatives of this class have done about the relation of religion to education. The very government which revived religious education in our schools has done nothing in almost twenty years to see that there are trained teachers to carry out its program. The implication is that in questions of religion no careful training is necessary. Indeed our educated bureaucratic bourgeoisie seems to be divided between a minority (at the lower levels of power) who would substitute the religion of progress entirely for our religious past, and a majority who have some continuing traditional relationship with the religion of the past in a very attentuated form.

It is here that the already described qualification of the ancient view of religion and the state comes into play. To repeat that qualification: unassisted reason is able to know that without religious beliefs and actions no society whatever can last, but reason is unable to determine which should be the particular public religion. This ancient position indeed seems to be embodied in the opinions of our present rulers. Many of them see the necessity of religion, but they are not in a position to take seriously the claims of Christian revelations. That is, they accept Christianity as the best tradition, but not as true. They are indeed in a similar position to the Roman aristocrats who attempted to defend the traditional paganism against the inroads of the new religion in the first centuries of our era. Some of their cleverest members, especially those who have taken seriously their training in the natural or social sciences, opt for the new religion of progress. Two conservative influences are at work here however: (1) ruling classes have always had a suspicion of their most advanced and theoretically clever members, (2) even some of the ruling class who think Christianity complete nonsense are hesitant in any opposition, either because they are truly agnostic and therefore put forward no opposition to religion and/or because they see the need of religion for social cohesion.

Indeed our present situation is illuminated when one compares the position of the Roman Catholic Church with that of the large non-Roman churches. The Church of Rome can insist on the maintenance of its right that its members are educated in a Catholic ethos, because sufficient of its laymen support the authority of its clerics in the belief that the Christian revelation is central to the understanding of human existence, and that therefore revelation must be the binding force that holds the educational process in unity. It is indeed remarkable to compare the present situation, vis-à-vis education, of the Roman Church and the non-Roman churches. In October 1962 the Roman Catholic bishops of Ontario suggested to the government (along with other matters) that the ecclesiastical control of the education of Roman Catholic children should be extended to the high school level and that for its proper maintenance special Catholic Teachers Colleges should be established. This would round out the whole system of Catholic education at all ages. Compare this with the situation of the non-Roman churches, which find themselves in retreat from participation in the educational process at all levels. Church colleges hardly do more than keep their nominally Christian structure, etc., etc. This comparison is not made to express agreement with the Roman Catholic view of education or authority but simply to accentuate the obvious fact that the Roman Catholic hierarchy can press for extension at this time only because they carry the majority of their laymen with them, and they do this because a sufficient number of laymen consider Christianity true and therefore essential to education. To put it bluntly, Roman Catholic laymen will use the vote to effect their religious interests because they take those religious interests seriously. Protestantism is still a voting force to be reckoned with, but largely only in an anti-Catholic and predominantly rural form.

If such an analysis be even comparatively accurate, two questions must be asked of those Christian leaders who defend strongly the continuance of Christian teaching in our state schools: (a) has the Christian Church the right to maintain that teaching? and (b) if it has the right, is it wise to insist on it? To the first of these questions only a qualified answer can be given. The second, in my opinion, can be answered negatively with some force.

The difficulty of answering question (a) in a more qualified way is the confusion of belief in our present situation. As I have said earlier, the right of a particular religion to public status (within the conservative account of the matter) is a right of tradition, that is, a right based on the dominant ethos of a particular society. In our present situation the dominant ethos is extremely confused and therefore the right is difficult of definition. On the one hand, it is simply outdated to say that the dominant ethos of Ontario is Christian and therefore to grant as self-evident (as some do) the right of the state to teach that religion in the schools. On the other hand, it is equally foolish to deny the continuing power of the non-Roman Christian tradition in Ontario. I find the argument of those liberals who say that large numbers of non-Roman Christians in our society have no rights vis-à-vis their religion in the state schools simply not understandable. I think the following evaluation can fairly be made: in the newer and more highly technologized parts of Ontario, the fact of pluralism exists and therefore the non-Roman churches have little right to public status for their religion. In the more traditional parts of Ontario, they have a real right to public status.

This last generalization implies why it seems to me unwise of Christian leaders to insist too strongly on their rights in the state schools. It is unwise because inevitably our society is becoming more and more technologized (if we exclude the possibility of catastrophe, the likelihood of which we cannot predict). Therefore, the increasingly dominant portions of our society will be those where the fact of pluralism must be unequivocally recognized. In other words, the traditionalist Christianity is going to be a wasting asset. The Church has always been too wise to rely on wasting assets. This loyalty to tradition is a wasting asset particularly in a state capitalism such as ours which is run by the bureaucratic élites in corporations and government. For by its very structure state capitalist bureaucracy teaches men that loyalty to anything other than a particular definition of efficiency is not a virtue. Bureaucracy such as ours is therefore by definition an environment where traditional loyalties are destroyed. The bureaucratic élites in our corporations have shown what little loyalty they feel for our nation. Is it likely that they will show it to the traditional religion?

In such a situation the danger that the churches face is of finding themselves used by the state, and, because of that using, of compromising their ability to proclaim the supernatural Gospel to the young. States change (the modern word is *develop*). It is surely a likely bet that the educational policies of the state in Ontario will be increasingly secularized, outside the Roman Catholic sector. The danger of being used becomes the following. At the present moment the dominant elements in the state still want the teaching of the Christian religion in the schools, even while it is divided against itself by using the religion of humanity and progress as the cement at deeper levels of the educational process. In other words, the state wants religion in the schools but does not want it there seriously. In such a situation Christianity is in the schools more and more as a façade of tradition. But the façade serves the passing interest of the state without really serving the interests of the churches. When the state has become secularized, it will quickly free itself of its use of the church. The religion of humanity and progress will reign monolithically in the schools. In the meantime the churches may have persuaded themselves that their educational interests are sufficiently served by the maintenance of this façade of public education. In accepting the present superficial system it prejudices its case with the young because it says by implication that the demand of supernatural truth upon their intellects is limited to a few thin platitudes.

The argument that it is not wise of church members to defend with intensity the present system does not however settle the question of the need for a public religion. Indeed there is only one group in our society to whom this question is easy of solution, namely the believers in the religion of progress, mastery and power. Assuming their religion to be self-evidently true to all men of good will, they are forceful in advocating that it should be the public religion. They work for the coming of the universal and homogeneous state with enthusiasm; they await its coming with expectation. Such a belief, of course, appears nonsense to those of us (Christian or non-Christian) who hold the conservative principle that belief in a "higher" divine power is a minimum public necessity if there is to be constitutional government. And it seems nonsense not only on the basis of what has been said about the matter in the traditional political philosophy, but also because of the evidence of the nature of the most advanced industrial societies. Has the secular state, and the religion of progress which dominates its education, led to widespread happiness in North America in the last forty years? How can we escape the fact that the necessary end product of the religion of progress is not hope, but a society of existentialists who know themselves in their own self-consciousness, but know the world entirely as despair? In other words, when the religion of progress becomes the public religion we cannot look forward to a vital religious pluralism, but to a monism of meaninglessness. And what becomes of the constitutional state in a society where more and more persons face their own existing as meaningless? Surely the basic problem of our society is the problem of individuals finding meaning to their existence. The most important cause of the psycho-pathological phenomena, which are becoming terrifyingly widespread at all échelons in North America, is just that human beings can find no meaning to their existence. Neither the Freudian nor the Marxian descriptions or therapies can account for or cure these new psychopathologies. The religion of progress may have been able to kill Christianity in the consciousnesses of many, but it has not succeeded in substituting any other lasting system of meaning. In such a situation the question of the public religion becomes crucial in western society. Marxism has proved a successful public religion in the U.S.S.R., but it will not adequately fulfil this function for more than another generation, as it is already being openly assailed by existentialist criticism and will not be able to withstand that criticism. The attempt by such men as Dr. Erich Fromm to make a humane Marxism the religion of North America will surely be also broken by the existentialist criticism.

It will, of course, seem unfair to the exponents of secularism that I have called what they advocate a religion. They will deem it unfair because they think that what they advocate is a product of reason alone and therefore should be called philosophy and not religion. What indeed buttresses their belief on this matter is one of their own central assumptions: namely that in the modern age philosophy is going to fulfill the functions that in the past were fulfilled by both philosophy and religion. This assumption is itself part of the religion of progress and is denied by those (such as myself) who take their philosophy and religion from the older tradition. The older tradition says that philosophy and religion fulfill different (albeit related) roles in the lives of human beings and that the practice of both are necessary to the healthy life of a society. It says that not many men will become philosophers; but that all men are inevitably religious. It is on these principles that one is forced to distinguish (even when its proponents do not) between modern philosophy and the modern religion (namely that of progress). Another implication of this traditional principle is that religion has a more direct relation to the public sphere than has philosophy. When Plato first used the word theology he used it to describe an activity which was public in a way that philosophy was not.

It is this distinction between religion and philosophy which leads those who accept it (whether Christian or non-Christian) to their dilemma in the present practical situation. On the one side of the dilemma there is the need of some public religion which is more than the ever-increasing externalizing of ourselves by the religion of progress. On the other side of the dilemma there is the fact that no religion at the moment can be taken unequivocally as the publicly received tradition of our society. Because of this situation, the proper claims of pluralism must be met in any constitutional state. In short, the dilemma is: there is need of a public religion, yet it is quite unclear what that public religion should be. It is the difficulty of this dilemma which makes those who grasp only one side of it so intransigent in their advocacies.

Perhaps it is not possible to reconcile these claims at this stage of our era. If this should sadly be the case then men who see the religious problem seriously will be forced to retreat from the public sphere and concern themselves simply with what they consider to be the true religion. The public sphere will then be turned over to the advocates of the religion of progress and mastery, to do what they want with it. This seems likely to be a necessity, but it can only be a sad one. For the surrender of this attempt at reconciliation not only implies an admission of the impotence of human charity, but also a total admission of the barren future of our civilization.

Canadian Fate
and
Imperialism

TO USE THE LANGUAGE OF FATE is to assert that all human beings come into a world they did not choose and live their lives within a universe they did not make. If one speaks in this way, one is often accused either of being pessimistic or of holding a tragic view of life. Neither of these accusations is correct. To say that one holds a tragic view of life would be to follow Nietzsche in thinking that Dionysian tragedy was a higher stance than that of Socrates; I do not think this. And the words optimistic and pessimistic are surely most accurately used, following Leibniz, to describe what one thinks about the nature of things, whether the world is good or not. It is quite possible to use the word "fate," and to think that "nature" is good, and not contradict oneself. It is in my opinion a sensible way to talk about events, though obviously it is far from the liberal dogmas within which most people are taught to think.

A central aspect of the fate of being a Canadian is that our very existing has at all times been bound up with the interplay of various world empires. One can better understand what it is to be Canadian if one understands that interplay. As no serious person is interested in history simply as antiquarianism but only as it illumines one's search for the good in the here and now, let me set the problem in its most contemporary form — Vietnam. What our fate is today becomes most evident in the light of Vietnam. It is clear that in that country the American empire has been demolishing a people, rather than allowing them to live outside the American orbit.

The Americans are forced to that ferocious demolition because they have chosen to draw the line against the Chinese empire in a country

where nationalism and communism have been in large measure identified. How does this affect Canadians? On the one hand, many Canadians,
whether their moral traditions come from Judaism, Christianity, the
liberal enlightenment, or a mixture, are not yet so empty that they can
take lightly the destruction of a people — even in the case of Asians. On
the other hand, the vast majority of Canadians are a product of western
civilization and live entirely within the forms and assumptions of that
enterprise. Today the enterprise of western civilization finds its spear-
head in the American empire. In that sense our very lives are inevitably
bound up in the meeting of that empire with the rest of the world, and
the movements of war which draw the limits in that meeting. The depth
of that common destiny with the Americans is shown in the fact that
many Canadians who are forced to admit the sheer evil of what is being
done in Vietnam say at the same time that we have no choice but to stand
with the Americans as the pillar of western civilization. Beyond this
kind of talk is of course the fact that this society is above all a machine
for greed, and our branch plant industry is making a packet out of the
demolition of Vietnam.

Our involvement is much deeper than the immediate profits of
particular wars. Our very form of life depends on our membership in
the western industrial empire which is centred in the U.S.A. and which
stretches out in its hegemony into parts of western Europe and which
controls South America and much of Africa and Asia. Somewhere in
the minds of nearly all Canadians there is the recognition that our
present form of life depends on our place as second class members of that
system. By "second class" I do not imply a low status, because there are
a large number of classes within it. It is much nicer to be a Canadian
than a Brazilian or a Venezuelan, or for that matter an Englishman.

Indeed our involvement in the American empire goes deeper than a
simple economic and political basis; it depends on the very faith that gives
meaning and purpose to the lives of western men. To most Canadians,
as public beings, the central cause of motion in their souls is the belief in
progress through technique, and that faith is identified with the power
and leadership of the English-speaking empire in the world.

This then is why our present fate can be seen with such clarity in
the glaring light of Vietnam. The very substance of our lives is bound

up with the western empire and its destiny, just at a time when that
empire uses increasingly ferocious means to maintain its hegemony.
The earlier catastrophes and mass crimes of the age of progress could
be interpreted as originating entirely with other peoples, the Germans, or the Russians. They could be seen as the perverse products
of western ideology — national socialism or communism. This can
no longer be said. What is being done in Vietnam is being done by
the English-speaking empire and in the name of liberal democracy.

NOT ONLY IN OUR PRESENT but in our origins, Canada was made
by western empires. We were a product of two north-western empires
as they moved out in that strange expansion of Europe around the
world. It is essential to emphasize that they were north-western.
Hegel's language is here the clearest. He speaks of the "germanische
Geist," and in using those words he does not mean the German spirit.
He means geographically those European lands whose rivers flow into
the North Atlantic. He means the particular secularizing Christianity
which characterized those lands. He understands that the dominant
spirit of the modern age is no longer in the Mediterranean peoples,
but has passed northward and westward to the Abendland.

If one is to understand Canada one must understand the history of
those empires — and not simply in terms of what they did, but in terms
of the spirit which drove them to such enormous motion. If one is to
pick the society where modernity first makes its appearance in a more
than individual way, one must pick England. To understand English
modernity one must look above all at that unique meeting of Calvinist Protestantism and the new secular spirit of the Renaissance. That
secular spirit can be seen in the new physical science whose origins we
identify with Galileo, and in the new moral science of Machiavelli. It
was the liberals' superficial interpretation of what we call the Renaissance to see such thinkers as a return to the Greeks, when they were
a profound turning away from the ancients. The role of Calvinism in
making possible the capitalism which has shaped the western world

has been described by Weber.[1] He sees with great clarity how Calvinism provided the necessary ethic for capitalism; what he does not understand is that deeper movement of the mind in which the Puritans were open to the new physical and moral science in a way the older Christianity was not. You can see this acceptance taking place in the seventeenth century. At the end of the sixteenth century, Shakespeare writes: "to set the murderous Machiavelli to school." But during the seventeenth century, Bacon, Hobbes, and Locke have achieved the terrible task of making Machiavelli widely respectable, and the new secular, moral and physical science is particularly welcomed by the Protestants. The union of the new secularism and Protestantism brought forth the first great wave of social modernity in England and its empire.

These days when we are told in North America that capitalism is conservative, we should remember that capitalism was the great dissolvent of the traditional virtues and that its greatest philosophers, Hobbes and Locke, Smith and Hume, were Britishers. In the appeal to capitalism as the tradition it is forgotten that the capitalist philosophers dissolved all ideas of the sacred as standing in the way of the emancipation of greed. For example, the criticism of any knowable teleology by Hume not only helped to liberate men to the new natural science, but also liberated them from knowledge of any purposes which transcended the economically rational. It is not surprising that North America was won by the English empire rather than the French. It is enough to read John Nef's book about the differing uses of iron in England and France in the seventeenth century. Despite the work of Henri IV, Richelieu, and Colbert, France was not to the same degree an initiator of capitalism and modernity.

The French who were left as an enclave on the shores of the St. Lawrence came from an earlier tradition, before France had initiated the second great wave of modernity with Rousseau and the French Revolution. What is so endearing about the young French-Canadians revolting against their tradition is that they sometimes write as if Voltaire's

---

1 This question is discussed at greater length in the first essay of this volume. Prof. G. E. Wilson has rightly pointed out to me that in emphasizing the importance of the English in this history, I have underestimated the importance of the Dutch.

Candide had come off the press last week instead of two hundred years ago. One's enchantment is however limited by the knowledge that their awakening to modernity, which seems to them an expression of independence, in fact leaves them wide open to conquest by a modernity which at its very heart is destructive of indigenous traditions. Of course, many stylish French-Canadian liberals are quite clear that their espousal of the modern does not consistently include any serious interest in the continuance of their own traditions, including even language.

Although the English who conquered North America were of a more modern tradition than the French left in their enclave, it must be remembered that there was always a strong losing party in all the great public events in which modernity put its stamp on English society. Progressivist historians do not write much about the losers of history, because belief in progress often implies the base assumption that to lose is to have failed to grasp the evolving truth. Nevertheless, the losers existed and they are worth reading now that we see what kind of society the winners have made. We can read what Hooker wrote against the Puritans and the society they would build. Above all, the views of the losing party can be found in the greatest of English prose stylists. Swift was a comic genius because he understood with clarity that the victory of the Whigs was not simply a passing political event, but involved new intellectual assumptions. In the quarrel between the ancients and the moderns, Swift knew why he accepted the ancients against the new moral science of Hobbes and Locke.

Though the empire of the English was the chief of the early driving forces towards modernity, many traditions from before the age of progress remained alive in parts of English society, and some of these existed in an inchoate way in the early English-speaking peoples of this country. It would be balderdash to imply that the early English-speaking leaders of Canada had a firm tradition like their French compatriots, or that they were in any sense people who resisted modernity with the clarity of Swift. Most of the educated among the Loyalists were that extraordinary concoction, straight Locke with a dash of Anglicanism. They were above all a product of the English empire, and the victory of modernity had long since been decided in favour of the Whigs. What can be fairly said, however, is that they were not so given over to modernity as were the leaders of the

U.S., particularly in so far as the Americans had incorporated in their revolution a mixture of Locke with elements of Rousseau. The fact that the Canadians had consciously refused the break with their particular past meant that they had some roots with tradition, even though that tradition was the most modern in Europe up till the eighteenth century. Indeed, when one reads the speeches of those founders whom we celebrated in 1967, one is aware of their continual suspicion of the foundations of the American republic, and of their desire to build a political society with a clearer and firmer doctrine of the common good than that at the heart of the liberal democracy to the south. (One would never know this from what one reads about our founders in our liberal text books.)

Nevertheless, having asserted these differences, what is far more important is to repeat that the English empire was a dominant source of modernity. The early Canadian settlers may have wanted to be different from the Americans in detail but not in any substantial way which questioned that modernity. I emphasize this for a personal reason. A couple of years ago I wrote a book about the dissolution of Canadian sovereignty. These days when psychologising is the chief method for neutralising disagreeable opinions, my psyche was interpreted as harking back in nostalgia to the British Empire and old-fashioned Canada. This was the explanation of why I did not think that the general tendencies of modern society were liable to produce human excellence. In this era when the homogenising power of technology is almost unlimited, I do regret the disappearance of indigenous traditions, including my own. It is true that no particularism can adequately incarnate the good. But is it not also true that only through some particular roots, however, partial, can human beings first grasp what is good and it is the juice of such roots which for most men sustain their partaking in a more universal good? Still, regret, however ironical, is not an adequate stance for living and is an impossible stance for philosophy. Conservatism is a practical stance; it must be transcended to reach philosophy. What I said in that book was that the belief that human excellence is promoted by the homogenising and universalising power of technology is the dominant doctrine of modern liberalism, and that that doctrine must undermine all particularisms and that English-speaking Canada as a particular is wide open to that doctrine.

IN THE NINETEENTH CENTURY the European empires modernized themselves. Nearly all those aspects of their cultures from before the age of progress disappeared. The fact that England came into that century with a vast empire (despite its loss of the American colonies), and as the pioneer of industrialism, meant that it started with an enormous advantage over the other modernizing empires. In some ways it was this sense of advantage and unquestioned power which made it indifferent to immediate political control in Canada. But after 1870 industrial and imperial competition was the order of the day and England threw itself into the wild scramble for more possessions and greater imperial control to counter the growing strength of its European rivals. The imperialism of the last half of the nineteenth century is modern man (man as Hobbes has said he is) realizing his potentialities. The culmination of that European process was the war of 1914.

Canada as always was involved in the general western fate. Just read how English-speaking Canadians from all areas and all economic classes went off to that war hopefully and honestly believing that they were thereby guaranteeing freedom and justice in the world. Loyalty to Britain and loyalty to liberal capitalist democracy was identified with loyalty to freedom and justice. For example, I have met people from Cape Breton who were so cut off from the general world in 1914 that they thought that Queen Victoria was still reigning and took for granted it was their duty to fight for her. When one thinks what that war was in fact being fought about, and the slaughter of decent men of decent motive which ensued, the imagination boggles. As that war spelled out the implicit violence of the West, it also spelled out Canadian fate.

First, it killed many of the best English-speaking Canadians and left the survivors cynical and tired. I once asked a man of that generation why it was that between the wars of 1914 and 1939 Canada was allowed to slip into the slough of despond in which its national hope was frittered away to the U.S. by Mackenzie King and the Liberal party. He answered graphically: "We had our guts shot away in France." The energy of that generation was drained away in that conflict so that those who returned did not have the vitality for public care, but retreated into the private world of money making. Canada's survival has always required the victory of political courage over immediate and individual economic advantage.

Secondly, English-speaking Canadians in the name of that brutal struggle between empires forced French-speaking Canadians to take part in a way which they knew not to be theirs. If Canada were to exist, English- and French-speaking peoples had to have sufficient trust to choose to be together rather than to be Americans. The forcing of the French by fanatics such as Sam Hughes and the culmination of that process in the election of 1917 meant that the French-Canadians saw themselves threatened more by English-speaking Canadians than by the deeper threat to the south. Mackenzie King's stand in the election of 1917 must be taken to his political credit, and God knows he needs credit somewhere. In saying that, however, one must remember that between the two great wars King and the Liberal party kept the flames of that hostility alive in Quebec so that they could take the full political benefit from it.

The third great effect of that war in Canada was due to the policies of the ruling classes in Great Britain. In the face of the competition from other European empires, the British ruling classes acted as if their only hope of continuing power was to put their fate into the hands of the American empire. That process is epitomized in the career of Winston Churchill. High rhetoric about partnership among the English-speaking peoples has been used about this process. It cannot, however, cover the fact that Great Britain's chief status in the world today is to do useful jobs for its masters and to be paid for so doing by the support of the pound and the freedom to provide entertainers and entertainment for the empire as a whole. The American empire may be having its difficulties with France and Germany, but it does not have them with Great Britain. Leaving aside the complex question of whether this status was the best that the English could achieve in the circumstances, it is clear that its effect on the possibility of Canada being a nation has been large. The elimination of Great Britain as an independent source of civilization in the English-speaking world greatly increased the pull of English-speaking Canadians to an identity with the centre of that world in the United States. It is an ambiguity of present Canada that some serious French Canadians now turn to France for support against the English-speaking technological sea. They so turn just as English-speaking Canadians can no longer turn to Great Britain for alternative forms of life to those which press from the south. This present turning is ambiguous because for so long English-speaking Canadians were told by French Canadians that we were not properly Canadian because of our connection with Great Britain. English-speaking Canadians now lean on similar criticism when the great general is welcomed in Quebec.

THE SUPREMACY OF THE AMERICAN EMPIRE in the western world was important for Canada not only in the geographic and economic senses that our nation had to try to exist in the very presence of the empire, but in the much profounder sense that the dominance of the United States is identified with the unequivocal victory of the progressive spirit in the West. The older empires had some residual traditions from before the age of progress — the French more, the British less. The United States is the only society that has none. The American supremacy is identified with the belief that questions of human good are to be solved by technology; that the most important human activity is the pursuit of those sciences which issue in the conquest of human and non-human nature. Anybody who is in university today, and knows where he is, knows both that these are the ends for which the university exists and that the universities are becoming almost the chief institutions of our system.

The gradual victory of the progressive spirit has taken place in interdependence with an enormous expansion of the western peoples into the rest of the world. The era of modern thought has been the era of western imperialism. Imperialism, like war, is coeval with human existence. But the increasingly externalised view of human life which is the very nature of the progressive spirit has given and will continue to give an enormous impetus to imperialism. As the classical philosophers said, man cannot help but imitate in action his vision of the nature of things. The dominant tendency of the western world has been to divide history from nature and to consider history as dynamic and nature controllable as externality. Therefore, modern men have been extremely violent in their dealings with other men and other beings. Liberal doctrine does not prepare us for this violence because of its identification of technology with evolution, and the identification of evolution with movement of the race to higher and higher morality. Such a doctrine could not understand that an expanding technological society is going to be an imperialist society even when it is run by governments who talk and sometimes act the language of welfare both domestically and internationally. Among advanced liberals, this failure was compounded by the naive account in Marxism of imperialism as simply a product of late capitalism. In the case of the American empire, the vulgarity of the analysis can be seen in the assessments of the presidency of F. D. Roosevelt. The "right" wing has accused him of being soft about American interests at Yalta; the "left" wing has seen him as a lover of humanity whose untimely death prevented him from stopping the Cold War and building a world based on the principles of the United Nations. In fact under his presidency the U.S. at last moved fully into the imperial role for which its dominant classes had been preparing since John Hay. Our modern way of looking at the world hides from us the reality of many political things; but about nothing is it more obscuring than the inevitable relation between dynamic technology and imperialism.

Of course, what has happened in our immediate era is that the non-western nations have taken on western means, both technical and ideological, as the only way to preserve themselves against the West. They now move from being simply the sufferers of Western dynamism to having an active imperialist role of their own. Indeed Russian and Chinese imperialism present an undoubted challenge to the West. There is equal distortion in the rhetoric of those who see the American empire as the sole source of violence in the world and in the rhetoric of those who see an essentially peaceful western world defending itself against communism. Modern imperialism — with all its ideological and technical resources — may have been invented in the West, but it is not now confined to it.

TO LIVE IN A WORLD of these violent empires, and in a satellite of the greatest of them, presents complex problems of morality. These problems may be stated thus. In human life there must always be place for love of the good and love of one's own. Love of the good is man's highest end, but it is of the nature of things that we come to know and to love what is good by first meeting it in that which is our own — this particular body, this family, these friends, this woman, this part of the world, this set of traditions, this country, this civilization. At the simplest level of one's own body, it is clear that one has to love it yet pass beyond concentration on it. To grow up properly is not to be alienated from one's own body; but an adult who does not pay reverence to anything beyond his own body is a narcissist, and not a full human being. In many parts of our lives the two loves need never be in conflict. In loving our friends we are also loving the good. But sometimes the conflict becomes open. An obvious case in our era is those Germans who had to oppose their own country in the name of the good. I have known many noble Jewish and Christian Germans who were torn apart because no country but Germany could really be their own, yet they could no longer love it because of their love of the good.

This is why the present happenings in Vietnam are particularly terrible for Canadians. What is being done there is being done by a society which is in some deep way our own. It is being done by a society which more than any other carries the destiny of the West, and Canadians belong inevitably to that destiny. Canada could only continue to be if we could hold some alternative social vision to that of the great republic. Yet such an alternative would have had to come out of the same stream — western culture. Indeed our failure to find such an alternative is bound up with the very homogenising path of western history. So we are left with the fact. As the U.S. becomes daily more our own, so does the Vietnam War.

The majority of North Americans do not seem to believe that love of their own and love of the good are exposed to stringent conflict in Vietnam. They assume that the structure of our society is essentially good, that it requires to be defended against aggression, and that it is against aggression that the American troops are engaged in Vietnam. They are either not much concerned with the actual history of the conflict, or else have been convinced by propaganda that there is a gallant country, South Vietnam, which is defending itself from aggression with American help. When a more explicit ideology is sought, the position becomes divided at one particular point. Are we to fear the Vietnamese and beyond them the Chinese because they are non-western, or because they are communist? Is it the old Europocentric fear of the Asian hordes under Asian tyranny as a threat to the freedom and right which belong essentially to the West? Or is it because the Asians have taken on communism that they are to be feared? It is not easy to hold these two positions together for the reason that Marxism is an advanced product of the West which appealed to British industry, French revolutionary ideas, and German philosophy. Of course many people in North America no longer appeal to any ideology beyond our affluence. They take the line that it's either them or us, and this position is wrapped up in Darwinian packaging which says that any means are permissible that allow us to protect our own.

For a minority, the events in Vietnam must help to push them over that great divide where one can no longer love one's own — where indeed it almost ceases to be one's own. Vietnam is a glaring searchlight exposing the very structure of the imperial society. Even if hopefully the violence there should ease off, the searchlight has still been cast on the structure. We can never be as we were, because what has been done has been done. Some could see the structure of that society before the last years, but Vietnam has been for many the means to a clearer analysis. It has had this result because here are obvious facts which cannot be accounted for within the usual liberal description by which the society is legitimised to its own members and to the world.

Many liberals who do not find the events in Vietnam easy to stomach sometimes talk as if what were happening there were some kind of accident — if only that Texan had not got into the White House, etc., etc. Such a way of thinking is worthy only of journalists. Let us suppose that the American ruling class (through either of its political instruments) comes to see more clearly what a tactical error it has made in Vietnam and allows the war to tail off. It still governs the most powerful empire in the history of the world. It may learn to carry out its policies (e.g., in South America) more effectively and without such open brutalities. But it will have to have its Vietnams if the occasions demand, and we will have to be part of them. A profounder liberal criticism is made by those who say that the health of the western empire is shown by the extent of dissent against war. They maintain that only the traditions of the West make such dissent possible and that that possibility shows us the essential goodness of liberal society. This argument turns on a judgement of fact — an extremely difficult one. Does this dissent in the West present a real alternative of action, or is it simply froth on the surface which is necessary to the system itself as a safety valve? I am not sure. I lean to the position that dissent on major questions of policy is impotent and that the western system has in truth achieved what Michels called "the bureaucratising of dissent." [2]

The word "alienation" has become a cliché to be thrown about in journalistic chitchat. Surely the deepest alienation must be when the civilization one inhabits no longer claims one's loyalty. It is a rational alienation, and therefore not to be overcome by opting out of the system through such methods as LSD and speed. The ecstasy therein offered is just another package which one buys from the system and which keeps people quiet. Indeed the depth of the alienation is seen in the ambiguity of the words "one's own." To repeat, the events in Vietnam push one towards that divide where one can no longer love one's own — to the point where the civilization almost ceases to be one's own. Yet it is impossible to give up the word "almost." Think of being the parent or the child of a concentration camp guard. One would want to say: "This person is not my own," and yet one could not. The facts of birth are inescapable. So are the very facts of belonging to the civilization that has made one. It is this inevitability which leads to the degree of alienation and disgust which some people feel in the present situation.

There is a distinction between those who in their alienation find political hope in loyalty to one of the other empires, and those who cannot find such an alternative. Sartre, for example, takes part in politics, but

---

2 Since this was written some of the evidence is in. Clearly the dissent over the war in the U.S. had some effect on some decisions. It was not only the difficulty of winning the war which convinced the ruling class that the enterprise was a mistake, but also that too many of the influential young were being alienated from the purposes of the society by it. The dissent made this clear. No ruling class could afford to neglect such a circumstance particularly when its society was faced at the same time by a major racial crisis. But it would be wrong to carry the consequences too far. The Democratic party was punished by losing the presidency — but not by the dissenters, rather by the settled and unsettled bourgeois. Dissent was able to expose the folly of defending imperial interests in such a misguided way in Vietnam. But it cannot be effective in turning the U.S. from its course as a great imperial power.

he takes part in it as seeking the victory of the eastern empires, and is able to do so because of the freedom granted him in the western world. Which of these two positions is more adequate turns on the following question of fact: whether the qualities common to technological empires are more fundamental than their differences. This is not the place to discuss that extremely difficult question. It is clear, however, that the person who says "no" to his own and cannot substitute a hope in some other empire is in a position of much profounder alienation than those who can put their political trust elsewhere.

A similar difference in alienation will be found between those who place expectation in changing our society by reform or revolution and those who do not. Those, such as myself, who think that the drive for radical change in this society tends only to harden the very directions the society is already taking, and who think of the source of revolutionary fervour as arising finally from a further extension of the very modernity which has brought us where we are, inevitably find it more difficult to know how to live in this society than those who have expectations from radical activity.

Some people, particularly some of the young, will say that I have used a lot of words about the obvious. They may say that they have known since they began to think that this society is quite absurd and that sanity requires one to be either indifferent or hostile to it. Why write so long about what is so evident? However, finding that one is hostile or indifferent to a society may be a necessary discovery, but it is always an emasculating one. Man is by nature a political animal and to know that citizenship is an impossibility is to be cut off from one of the highest forms of life. To retreat from loyalty to one's own has the exhilaration of rebellion, but rebellion cannot be the basis for a whole life. Like all civilizations the West is based on a great religion — the religion of progress. This is the belief that the conquest of human and non-human nature will give existence meaning. Western civilization is now universal so that this religion is nearly everywhere dominant. To question the dominant world religion is indeed to invite an alienation far greater than the simply political.

Nothing here written implies that the increasingly difficult job of preserving what is left of Canadian sovereignty is not worth the efforts of practical men. However disgraceful has been our complicity in the Vietnam War, however disgusting the wealth we have made from munitions for that war, one must still be glad that Canadian forces are not fighting there. This is due to what little sovereignty we still possess. So equally our non-involvement in the imperial adventures elsewhere will continue to depend on the possible maintenance of some waning sovereignty. But what lies behind the small practical question of Canadian nationalism is the larger context of the fate of western civilization. By that fate I mean not merely the relations of our massive empire to the rest of the world, but even more the kind of existence which is becoming universal in advanced technological societies. What is worth doing in the midst of this barren twilight is the incredibly difficult question.

Tyranny and Wisdom

INTRODUCTION

THIS ESSAY ATTEMPTS TO INTRODUCE *what is for me the most important controversy in contemporary political philosophy. Both the controversy and my comments on it may seem over festooned with the trappings of scholarship; but through them a question of magnitude arises. There is every reason to be suspicious of the trappings of scholarship these days. There is nothing phonier in our present universities than the exaltation of scholarship as if it were an end in itself. To be neuter before the question of good leads to that boasted neutrality in the multiversity which denies itself in its service of the modern state. Antiquarianism in the humanities has often been a means to cloak the fact that these studies have nothing significant to say about living in the technological era. Scholarship has, of course, always been a technique — a means through which men could come into the presence of the most serious questions. But when the thought that there are such questions has become dim in the positivist night, then scholarship becomes a technique serving no purpose beyond itself, and supposedly justifying its own existence. Research then becomes little more than an excuse for avoiding the ardours of teaching, and Walter Rathenau's crack appears true: "There are no specialists, only vested interests."*

This debasement of scholarship may turn people away from its benefits. The centre of true scholarship is the careful reading of what the wisest men have written about the most important questions. Both Strauss and Kojève have been scholars in that sense, and through that have moved beyond scholarship to that for which it is a means — wisdom. The question they raise in their controversy is whether the universal and homogeneous state is the best social order. This could only be discussed profoundly by those who had read what the wisest men in the past had written of tyranny and wisdom.

PROFESSOR LEO STRAUSS'S BOOK *On Tyranny* was published in 1948.[1] In 1954 a French translation was published with an accompanying essay by Monsieur Alexandre Kojève, entitled *Tyrannie et Sagesse,* and a reply to Kojève by Strauss.[2] In 1959 Strauss included his reply to Kojève in English in *What Is Political Philosophy?*[3]

My purpose in writing is not to give a summary of the controversy between Strauss and Kojève. Both men know better than I do what words are necessary to make clear what they mean. Modern academic writing is strewn with impertinent *précis* written by those who think they can say in fewer words what wiser men than they have said in more. In this paper many of Strauss's and Kojève's arguments will not be mentioned. Rather I intend to comment on certain propositions and arguments in that controversy which interest me because they appear to be fundamental to political theory. Nevertheless it is inevitable that I start by stating what the controversy in its most general form would seem to be about.[4]

1 L. Strauss, *On Tyranny—An Interpretation of Xenophon's* Hiero (Glencoe, Ill.: The Free Press, 1948).

2 L. Strauss, *De La Tyrannie,* Les Essais LXIX (Paris: Gallimard, 1954). Strauss' essay in this volume also includes his reply to Prof. Eric Voegelin's criticism of his original work. His controversy with Voegelin is, however, not my concern in this paper.

3 L. Strauss, *What Is Political Philosophy?* (Glencoe: The Free Press, 1959). In the English version of the essay, Strauss has cut out certain passages included in the earlier French version. These deletions do not radically change the English version. I regret them, however, because their inclusion in the French argument does bring out some of the implications in the controversy.

4 What an author has written at one place is inevitably illumined by what he has written at another. It is therefore impossible to write of Strauss or Kojève without having in mind their other writings. Nevertheless in this paper I will stick as much as possible to their writing about *Hiero.* When their doctrine about a particular matter is taken from elsewhere, this will be made clear in the footnotes.

Strauss affirms that "Tyranny is a danger coeval with political life."[5] He maintains that modern social science has not been able to comprehend modern tyranny. He recognizes that there are differences between antique and modern tyranny. "Present day tyranny, in contradistinction to classical tyranny, is based on the unlimited progress in the 'conquest of nature' which is made possible by modern science, as well as on the popularization or diffusion of philosophic or scientific knowledge."[6] Nevertheless to understand modern tyranny it is first necessary to understand "the elementary and in a sense natural form of tyranny which is pre-modem tyranny."[7] *Hiero* is "the only writing of the classical period which is explicitly devoted to the discussion of tyranny and its implications, and to nothing else."[8] Strauss's primary purpose is therefore to write a proper commentary on that work. By proper I mean that he is concerned with explicating what Xenophon wrote, not with his own thoughts about it. But it is clear that in doing this he affirms by implication that classical social science can understand tyranny in a way that modern social science cannot. It is this assumption in Strauss's commentary which Kojève makes explicit in his essay. Strauss in his reply to Kojève confirms that this is his contention.

Kojève never argues with Strauss about his interpretation of Xenophon. He continually uses the term Xenophon-Strauss in a way which makes clear that Strauss has correctly interpreted Xenophon's doctrine. He also agrees with Strauss that contemporary social science does not understand tyranny and in particular the relation between tyranny and philosophy.[9] Kojève nevertheless rejects the classical solution to the definition of tyranny; indeed he rejects classical political science in

---

5 *On Tyranny*, p. 1.

6 *What Is Political Philosophy?*, p. 96.

7 *On Tyranny*, p. 2.

8 *Ibid*, p. 1.

9 When I use the adjective "modern" about political philosophy, it is used generally to describe the political philosophy of Europe since Machiavelli. When I use the adjective "contemporary" about social science I mean these studies as they are generally practiced in our day, particularly in North America. As this paper is concerned with the differences between Strauss and Kojève, it will not be concerned with their criticisms of contemporary social science.

general. In its place he affirms the truth of Hegel's political theory as being able to describe tyranny correctly and indeed all the major questions of political theory. [10] Strauss does not question that Kojève has interpreted Hegel correctly. I will therefore use the phrase Hegel-Kojève in this essay. The centre of the controversy is then whether classical political science or modern political science (as perfected in Hegel) can the better understand the relation between the tyrant and the wise man or indeed any of the basic political questions.

In stating that the issue of the controversy is between classical and Hegelian political science, one must avoid a possible misunderstanding which if entertained would prevent one from taking Strauss seriously. Strauss affirms both that present day tyranny is different from ancient tyranny and that classical social science understands tyranny in a way that no modern social science can. From such statements the obvious question arises: how could the ancients understand something which did not exist in their day? (Such a question will be particularly pressing for those social scientists whose intellectual outlook may be briefly described as "historicism." Indeed historicism may prevent such scholars from even reading Strauss's arguments.) The apparent inconsistency is resolved in this way: Strauss says that what distinguishes modern tyranny from ancient tyranny is the presence in the modern world of a science that issues in the conquest of nature and the belief in the possibility of the popularization of philosophy and science. Both these possibilities were known to the ancient philosophers. "But the classics rejected them as 'unnatural,' *i.e.*, as destructive of humanity. They did not dream of present day tyranny because they regarded its basic presuppositions as

---

10 As I write chiefly for English-speaking readers, it is necessary to state here that Kojève's Hegel is not the gentlemanly idealist of the nineteenth century who became the butt of the British "realists" in this century. To Kojève the essential work of Hegel is *The Phenomenology of Spirit*. His Hegel is atheist and his thought contains all the truth implicit in existentialism and Marxism. Since his lectures of the 1930s Kojève has exerted a profound influence on the contemporary existentialists in France. As the popular North American usage of "atheist" is unclear, I will use it in this paper in the sense that Kojève used it in his *Introduction*. I am not certain whether Kojève's interpretation of Hegel is correct; but I am quite certain that Kojève's Hegel is incomparably nearer to the original than such English interpretations as those of Caird, Bosanquet, and Russell.

so preposterous that they turned their imaginations in entirely different directions."[11] In terms of this historical assertion, both Strauss's
affirmations can be made. Classical political science was not familiar
with modern tyranny, but it was familiar with the assumptions which
distinguish it from antique tyranny. Strauss is obviously asserting the
classical view that tyranny is a form of government common to all ages
and that the political philosopher can have knowledge of what is common
to these governments of disparate ages so that he can correctly call each
of them tyrannies. The contemporary social scientist may criticize such
a position, but his disagreement is a philosophical one. It does not arise
from an obvious mistake about the facts on Strauss's part.

It is difficult to know where to plunge into the controversy. Such
uncertainty is inevitable once the incipient political philosopher has
recognized that his study cannot avoid being metaphysical and that
therefore he must try to learn from those who can think more widely and
consistently than himself. For whatever else may be said about the philosophers who related their doctrine on political matters to their desire
to have knowledge of the whole, among the best of them there has been
a monumental consistency which related their doctrine on one issue to
what they taught on all others. Both Strauss and Kojève have studied
the masters with great care and therefore know in detail the political
teachings of the metaphysicians. They are both aware of the wide extent
to which the difference between classical and Hegelian political teaching
involves a difference of doctrine on nearly every major issue of political
theory. Indeed to state the obvious, the controversy implies throughout
a difference of opinion about the object and method of philosophy, in the
more than political sense of that word.[12] I do not want simply to check off
these differences in detail, yet on the other hand I do not feel competent
to define the central principles which divide classical from Hegelian
metaphysics. For these reasons I take the plunge into the controversy at
the point of a concrete political teaching.

---

11 *What Is Political Philosophy?*, p. 96.

12 It is interesting that the paragraph at the end of Strauss's reply to Kojève, in
which he describes the difference between the classical and Hegelian accounts
of the nature of philosophy, is not included in the English version. Perhaps it is
not too rash to infer that Strauss did not include it because of the general lack of
interest in metaphysical questions among English-speaking intellectuals.

Kojève affirms that the universal and homogeneous state is the best social order and that mankind advances to the establishment of such an order. The proof of this is according to Kojève found in Hegel's *Phenomenology of Spirit,* and Kojève sketches the argument of that book in a few pages of his essay.[13] From that sketch his account of the proper relation between tyranny and wisdom emerges. Alexander, pupil of a philosopher, was the first ruler who met with success in realizing a universal state, that is, an empire of which men could become citizens not simply because of their common ethnic or geographic background, but because they shared a common "essence." And that essence was, in the last analysis, their sharing of what modern men call "civilization" — the culture of the Greeks, which was for Alexander the culture of reason itself. But according to Hegel-Kojève, Alexander, as a Greek philosopher, could not overcome the distinction between masters and slaves. Thus his universal state could not be homogeneous — a society without classes. The distinction between master and slave was only overcome as a consequence of the Semitic religions (Judaism, Christianity, and Islam). For the West this culminated in Saint Paul. To quote:

"It is the idea of the *fundamental equality* of all those who believe in a single God. This transcendental conception of social equality differs radically from the Socratic-Platonic conception of the identity of beings having the same *immanent* 'essence.' For Alexander, a disciple of the Greek philosophers, the Hellene and the Barbarian have the same title to political citizenship in the Empire, to the extent that they HAVE the same human (moreover, rational, logical, discursive) 'nature' (=essence, idea, form, etc.) or are 'essentially' *identified* with each other as the result of a direct (='immediate') 'mixture' of their innate qualities (realized by means of biological union). For St. Paul there is no 'essential' (irreduceable) difference between the Greek and the Jew because they both can BECOME Christians, and this not by 'mixing' their Greek and Jewish 'qualities,' but by *negating* them both and 'synthesizing' them in and by this very negation into a homogeneous unity

13   *Les Essais*, pp. 266–280.

not innate or given, but (freely) *created* by 'conversion.' Because
of the *negating* character of the Christian 'synthesis,' there
are no longer any incompatible 'qualities,' or 'contradictory'
(=mutually exclusive), 'qualities.' For Alexander, a Greek philosopher, there was no possible mixture of Masters and Slaves,
for they were 'opposites.' Thus his *universal* State, which did
away with *race*, could not be *homogeneous* in the sense that it
would equally do away with 'class.' For St. Paul on the contrary,
the negation (*active* to the extent that 'faith' is an *act*, being
'dead' without 'acts') of the opposition between pagan Mastery
and Servitude could engender an 'essentially' *new* Christian
unity (which is, moreover, active or acting, or 'emotional,' and
not purely rational or discursive, that is 'logical') which could
serve as the basis not only for political *universality*, but also
for the social *homogeneity* of the State."[14]

The union of the political ideas of universality and homogeneity could not however result in the universal and homogeneous state
becoming a realisable political end when it came into the West as part
of Christian theism. That religion did not suppose such a state to be
fully realisable in the world, but only in the beyond, in the kingdom
of heaven. Homogeneity based on faith in a transcendent God could
only lead to the conception of the universal and homogeneous Church,
not to a universal and homogeneous state. For the universal and homogeneous state to be a realisable political end, Christian theism had first
to be negated. This negation was the accomplishment of modern
philosophy — an accomplishment of such men as Hobbes and Spinoza
which was completed by Hegel. Modern philosophy was able to secu-
larize (that is to rationalize or transform into coherent discourse) the
religious idea of the equality of all men. Thus the idea of the classless
society, then, is a derivative of the Christian religion because modern
philosophy in negating the Christian religion was aware of the truth
present in that which it negated. Thus the universal and homogeneous

---

14 *Ibid.*, pp. 273–274. The English translation of Kojève's essay has been
made by Michael Gold. This translation has now been published with
Strauss's writings on the matter. (L. Strauss, *On Tyranny* revised and
enlarged, Glencoe: The Free Press, 1963.)

state became a realisable political order (because of modern philosophy) and has been, is, and will be made actual by rulers.

Indeed whatever else should be said of Kojève's sketch of western political history, it is surely accurate to affirm that the universal and homogeneous state has been the dominant ideal in recent modern political thought, not only among those who have recognized their debt to Hegel but among many who would scorn Hegel's philosophy. Indeed the drive to the universal and homogeneous state remains the dominant ethical 'ideal' to which our contemporary society appeals for meaning in its activity. In its terms our society legitimises itself to itself. Therefore any contemporary man must try to think the truth of this core of political liberalism, if he is to know what it is to live in his world. The need to think the truth of this ideal remains even as its content empties into little more than the pursuit of technical expansion. It is Kojève's contention that Hegel's comprehension of the implications of asserting that the universal and homogeneous state is the best political ideal was complete in a way not present in any other political philosopher.

This sketch indicates the proper relation between politicians and philosophers, according to Hegel-Kojève. Only by discursive, philosophic reflection can a person become completely aware of the given historical situation at any moment. But the philosopher who is completely aware of the given historical situation must distinguish between that situation and the ideal. This distinction between the actual and the ideal leads to the negation of the given historical situation by struggle and work. This practical negating has always been accomplished by political tyrants. There would thus be no historical progress if philosophers did not instruct politicians in the meaning of actual historical situations.

Equally there would be no philosophical progress if practicing politicians did not realize the teaching of the philosophers in the world through work and struggle. This doctrine is, of course, of central significance in approaching the Hegelian dialectic. Philosophy is always the account of actuality as it has become in any particular epoch, including the contradictions of that epoch.[15] Therefore philosophy cannot hope to

---

15 It is quite impossible to describe here how the term contradiction can be applied to society as well as to thought in the Hegelian logic — namely the aspect

reach any conclusions which transcend the social situation of its age. Progress in philosophy, then, depends on the contradictions of a particular epoch being overcome by struggle and work, and this is always centred in the work of particular tyrants. Any other view of philosophy pre-supposes that philosophy is concerned with timeless concepts or, in other words, with an ahistorical eternal order. The belief that philosophy is concerned with the eternal order is based on the fundamental error of classical logic that being is eternally identical with itself. Kojève's motto could be the famous *tag* of Hegel's: "*Was die Zeit betrifft, so ist sie der daseiende Begriff selbst.*" The fundamental assumption of Hegelian logic (that being creates itself throughout the course of history and that eternity is the totality of all historical epochs) is only taken seriously at the level of politics in the recognition of the dependence of philosophers on the activity of tyrants. Kojève returns again and again in his writings to the point that Hegel alone has recognized fully the relation between the modern negation of theism and man's freedom to make the world (history). Only in the radical negation of theism is it possible to assert that there is progress — that is, that there is any sense or over-all direction to history. This progressive and atheist interpretation of Hegel illustrates and in turn is illustrated by the doctrine that progress in philosophy and the successful practice of wise tyrants depend upon each other.

One related facet of this interdependence between wisdom and tyranny is Kojève's assertion that the realization of the universal and homogeneous state will involve the end of philosophy. The love of wisdom will disappear because human beings will be able to achieve wisdom. For Kojève, this realization of wisdom has been first achieved in the writings of Hegel. But this appearance of the Sage had to be proceeded by the action of that tyrant (Napoleon) who established the basis of the universal and homogeneous state. According to Kojève, Hegel has produced the book (Bible) of wisdom which has definitely replaced the one with that

of Hegel's political teaching which has been most influential in Europe and at the same time most scorned among the philosophers of the English-speaking world. For a series of clear essays in which Kojève's account of Hegel is discussed see R. Queneau, (ed.) *Etudes Hegeliennes* (Neuchatel, 1955).

title which we have had for two thousand years.[16] This implies that for Kojève the events of 1830–1945 and after have been simply the completion in the world of that universal and homogeneous state which was initiated in one geographic area by Napoleon and which was completely understood by Hegel. Such an interpretation of Kojève would explain why he has used his philosophic talents simply to expound Hegel and why he has devoted most of his energies to the practical world.

Strauss's most general criticism of this account of western history is that it is based on the assumption that the universal and homogeneous state is the best social order. That is, according to Strauss, the Hegelian philosophy of history is essentially an attempt to interpret western history so that the two propositions "the universal and homogeneous state is the best social order" and "this social order will be built by man" will be shown to be true. Before proceeding to describe his arguments further, I must comment on what this general criticism by Strauss shows about the form of his argument as a whole. It is clear that in saying that Hegel's philosophy of history depends upon a universal proposition about all social orders, Strauss is speaking within the assumptions of classical philosophy; namely, that political philosophy stands or falls by its ability to transcend history, *i.e.*, by its ability to make statements about the best social order the truth of which is independent of changing historical epochs and which therefore cannot be deduced from any philosophy of history which makes positive statements about the historical process. But this assumption about political philosophy is in turn dependent on the assumption that the Socratic account of philosophy as a whole is true. Strauss also knows that for Hegel-Kojève the truth about the best social order is not prior to an interpretation of history and could not be known except at a certain epoch. This truth is reached by an argument which appeals to an interpretation of the sum of historical epochs which in totality, for the Hegelian, constitutes eternity. For the Hegelian, political philosophy does not stand or fall by its ability to transcend history, but rather by its ability to comprehend all history. Strauss knows that the difference between Hegel and the classics

---

16 *Les Essais*, p. 277.

about the place of "history" in the whole depends upon and illustrates a profound difference between them about the object, method, and standpoint of the study of philosophy in a more than political sense.

The question may then be asked: why does Strauss criticize Kojève from within an account of philosophy which Kojève does not accept? Why then does he not argue about the fundamental difference as to the nature of philosophy rather than the implications which follow from this difference concerning the proper relation between tyranny and wisdom? The answer to this clarifies the limits of Strauss's intentions. His first concern is not to refute Hegelianism but to show that the classical account of the relation between tyranny and wisdom is required by the classical account of philosophy; that is, that there is consistency between what the classics say about the whole and about politics, and that therefore classical political philosophy is not to be judged as a first phase of the subject, which has been left behind as mankind has progressed. Should this seem a limited task, it is perhaps worth pointing out that it is a necessary preliminary to the more difficult matter of being able to say that the political teaching of the classics is true in a way that modern political teaching is not. Also the need of this preliminary is particularly pressing for anyone who wishes to encourage the serious study of the classics in our era, when the doctrine of progress still influences the political beliefs of even the sophisticated. In such an era it is extremely difficult for men to contemplate the possibility that the political teachings of an ancient civilization could even be studied seriously, let alone in the light of the possibility that they are sounder than those of our day. [17]

Beyond this primary task, Strauss's secondary purpose is to show that the universal and homogeneous state, far from being the best social

---

17 That the first purpose of Strauss's argument is to stress the consistency between all aspects of classical philosophy might well have been made clearer. The difficulty of understanding his purpose is indeed increased for English readers by the fact that he does not include in the English edition of his work the last paragraph of the French edition, in which his purposes are beautifully described. Elsewhere in his writings Strauss has criticized historicism in its late nineteenth century form and shown the consequences of such a doctrine for the scope of political science. At no point in his writings has he, however, argued at length with Hegel's claim to have included history within metaphysics, and with the resulting relation between concepts and time.

order, will be (if realized) a tyranny, and therefore within classical assumptions destructive of humanity. Modern political philosophy, which has substituted freedom for virtue, has as its chief ideal (and an ideal which it considers realisable) a social order which is destructive of humanity. The relation of Strauss's second purpose to his first is evident. If the assumptions of modern political philosophy can be shown to lead to the dehumanizing of people, there is reason for scholars to take a more careful look at classical political philosophy than has generally been the case in the age of progress.

Strauss first argues that the universal and homogeneous state will not provide reasonable satisfaction for men, even on Hegelian assumptions about the objects in which men find the highest satisfaction. Strauss points out that, according to Hegel-Kojève, men find the highest satisfaction in universal recognition: "Men will have very good reasons for being dissatisfied with the universal and homogeneous state. To show this, I must have the recourse to Kojève's more extensive exposition in his *Introduction à la lecture de Hegel*. There are degrees of satisfaction. The satisfaction of the humble citizen, whose human dignity is universally recognized and who enjoys all opportunities that correspond to his humble capacities and achievements, is not comparable to the satisfaction of the Chief of State. Only the Chief of State is 'really satisfied.' He alone is 'truly free' (p. 146). Did Hegel not say something to the effect that the state in which one man is free is the Oriental despotic state? Is the universal and homogeneous state then merely a planetary Oriental despotism? However this may be, there is no guarantee that the incumbent Chief of State deserves his position to a higher degree than others. Those others then have a very good reason for dissatisfaction: a state which treats equal men unequally is not just. A change from the universal-homogeneous monarchy into a universal and homogeneous aristocracy would seem to be reasonable. But we cannot stop here. The universal and homogeneous state, being the synthesis of the Masters and the Slaves, is the state of the working warrior or of the war-waging worker. In fact, all its members are warrior workers (pp. 114, 146). But if the state is universal and homogeneous, 'wars and revolutions are henceforth impossible' (pp. 145, 561). Besides, work in the strict sense, namely the conquest or domestication of nature, is completed, for otherwise the universal and homogeneous state could not be the basis for wisdom (p. 301). Of course, work of a kind will still go on, but the citizens of the final state will work as little as possible, as Kojève notes with explicit reference to Marx (p. 435). To borrow an expression which someone used recently in the House of Lords on a similar occasion, the citizens of the final state are only so-called workers, workers by courtesy. 'There is no longer fight nor work. History has come to its end. There is nothing more to *do*' (pp. 114, 385). This end of History would be most exhilarating but for the fact that, according to Kojève, it is the participation in bloody political struggles as well as in real work or, generally expressed, the negating action, which raises man above the brutes (pp. 378n., 490–492, 560). The state through which man is said to become reasonably satisfied is, then, the state in which the basis of man's humanity withers away, or in which man loses his humanity." [18]

Strauss's criticism of Kojève then proceeds to a profounder level with the assumption of the classical philosophers that it is in thinking rather than in recognition that men find their fullest satisfaction. The highest good for man is wisdom. This being so, if there is to be satisfaction for all in the universal and homogeneous state, it must be that every human being is capable of systematic philosophical thought which alone, according to the classics, can lead to wisdom. Indeed Strauss affirms that the classical realization that only the few are capable of pursuing wisdom is central to the classical account of politics and indeed of human history as a whole.

"The classics thought that, owing to the weakness or dependence of human nature, universal happiness is impossible, and therefore they did not dream of a fulfillment of History and hence not of a meaning of History. They saw with their mind's eye a society within which that happiness of which human nature is capable would be possible in the highest degree: that society is the best regime. But because they saw how limited man's power is, they held that the actualisation of the best regime depends on chance. Modern man, dissatisfied with utopias and scorning them, has tried to find a guarantee for the actualisation of the best social order. In order to succeed, or rather in order to be able to believe that he could

---

18 *What Is Political Philosophy?*, pp. 128–129.

succeed, he had to lower the goal of man. One form in which this was done was to replace moral virtue by universal recognition, or to replace happiness by the satisfaction deriving from universal recognition. The classical solution is utopian in the sense that its actualisation is improbable. The modern solution is utopian in the sense that its actualisation is impossible. The classic solution supplies a stable standard by which to judge of any actual order. The modern solution eventually destroys the very idea of a standard that is independent of actual situations." [19]

Here again one sees the combination of Strauss's two purposes. The showing forth of the classical position as consistent combines with the description of modern theory as including assumptions which are destructive of humanity. Here also can be seen the complex of inter-related questions which need to be thought through if one is to judge about the main issue of the controversy.

The classical assumptions about the dependence of man make it clear that if the universal and homogeneous state were to be realized, it would be a tyranny and indeed the most appalling tyranny in the story of the race. Strauss ends his essay with a description of that tyranny:

"It seems reasonable to assume that only a few, if any, citizens of the universal and homogeneous state will be wise. But neither the wise men nor the philosophers will desire to rule. For this reason alone, to say nothing of others, the Chief of the universal and homogeneous state, or the Universal and Final Tyrant, will be an unwise man, as Kojève seems to take for granted. To retain his power, he will be forced to suppress every activity which might lead people into doubt of the essential soundness of the universal and homogeneous state: he must suppress philosophy as an attempt to corrupt the young. In particular he must in the interest of the homogeneity of his universal state forbid every teaching, every suggestion, that there are politically relevant natural differences among men which cannot be abolished or neutralized by progressing scientific technology. He must command his biologists to prove that every human being has, or will acquire, the capacity of becoming a philosopher or a tyrant. The philosophers in their turn will be forced to defend themselves or the cause of philosophy. They will be

---

19 *Ibid.*, pp. 131–132.

obliged, therefore, to try to act on the Tyrant. Everything seems to be a re-enactment of the age-old drama. But this time, the cause of philosophy is lost from the start. For the Final Tyrant presents himself as a philosopher, as the highest philosophic authority, as the supreme exegete of the only true philosophy, as the executor and hangman authorised by the only true philosophy. He claims therefore that he persecutes not philosophy but false philosophies. The experience is not altogether new for philosophers. If philosophers were confronted with claims of this kind in former ages, philosophy went underground. It accommodated itself in its explicit or exoteric teaching to the unfounded commands of rulers who believed they knew things which they did not know. Yet its very exoteric teaching undermined the commands or dogmas of the rulers in such a way as to guide the potential philosophers toward the eternal and unsolved problems. And since there was no universal state in existence, the philosophers could escape to other countries if life became unbearable in the tyrant's dominions. From the Universal Tyrant, however, there is no escape. Thanks to the conquest of nature and to the completely unabashed substitution of suspicion and terror for law, the Universal and Final Tyrant has at his disposal practically unlimited means for ferreting out, and for extinguishing, the most modest efforts in the direction of thought. Kojève would seem to be right although for the wrong reason: the coming of the universal and homogeneous state will be the end of philosophy on earth."[20]

COMMENTARY

I COMMENT ON TWO MATTERS alone in the tangle of questions which make up this controversy.

(A)   Strauss's assertion that the classical political philosophers considered the possibility of a science that issues in the conquest of nature as "'unnatural,' i.e., as destructive of humanity" and that therefore they turned their minds away from it, is clearly a proposition of consequence

---

20 *Ibid.*, pp. 132–133.

for political philosophy. Obviously it should also be a proposition of immediate practical interest for any person living in modern industrial society. This assertion demands answers to two questions: (i) Is this a true statement about the writings of the classical philosophers? and (ii) If it be true, were they right in so thinking?

In regard to the first question of fact Strauss gives three references in making the assertion. He asks the reader to compare Xenophon's *Memorabilia* Ii. 15 with Empedocles, fragment III. He cites Plato, *Theaetetus* 180c7–d5.[21] The inference which Strauss draws from Xenophon appears to me indubitable: that Socrates considered the practical application of physical philosophy to the control of nature to be "meddling" in a way that men should not and in a way that Empedocles had suggested that men could and ought. Nevertheless as soon as that is said, it must be emphasized that most modern scholars have not stated that the classical political philosophers turned their back on a science issuing in the conquest of nature. Presumably they do not draw the same inferences from these passages as does Strauss. Indeed it is clear that a matter of this consequence cannot be settled solely by an appeal to individual texts.

What is at stake in this question of fact can perhaps be clarified by stating Strauss's position in contradistinction to that of the Marxists. On one point of fact the Marxists and Strauss are in agreement: the classical philosophers did turn their backs on a science issuing in the conquest of nature. To the Marxists, the Greek philosophers turned their backs on technological advance because they wished to perpetuate the aristocratic society in which the majority of human beings served a minority through peasant and slave labour. The classical conception of philosophy and science, as the attempt to understand the eternal causes of all things, was the response of an aristocratic class desiring to perpetuate the social order most acceptable to itself. The theoretical standpoint of Greek science, admiring the contemplation of necessity, was related to the fear that the practical applications of science would destroy the privileged positions of those who were enabled to have leisure because of the work of the masses. But for the Marxists the liberation of all mankind from alienated labour is a goal that those who are so alienated will necessarily pursue. This

---

21 *Ibid.*, p. 96.

liberation can only be based on the domination of the realm of necessity by human freedom. Therefore the Platonic account of science held by aristocrats who had turned their backs on the domination of nature is patently inadequate, and the society in which it was prevalent was doomed to collapse. The basic Marxist assumption is that the domination of nature by human freedom is the indispensable condition of the liberation which all men seek. In short the Marxists agree with Strauss that the classical philosophers turned their backs on a technological science; they disagree about the Greeks' reason for so doing. The degree to which class interest was conscious or merely a product of "false consciousness" has largely been left unsettled by the Marxists.

In terms of this language of "the realms of necessity and of freedom" Strauss's position asserts an eternal and unchangeable order in which history takes its place and which is in no manner affected by the events of history. The realm of freedom is no more than a dependency of the realm of necessity. For Strauss the attempt to dominate the realm of necessity, far from being the condition of universal human satisfaction, implies the impossibility of true human excellence. He argues as follows: philosophy is the excellence of the soul. There cannot be philosophy in this sense unless there is an eternal and unchangeable order. But the belief that one can dominate the realm of necessity is to deny any eternal order which transcends history and in which history takes its place. Therefore the desire to dominate necessity leads to the denial of the possibility of human excellence.

This account of necessity and its relation to the pursuit of technological progress is also related to what I quoted earlier from Strauss: the classics did not consider universal human satisfaction possible because of human dependence and weakness. Also such an account of necessity is related to the very different account of human action found in the classics as compared to the moderns. Strauss refers to that classical account when he writes: "Modern man as little as pre-modern man can escape imitating nature as he understands nature." [22] The Kantian account of human freedom, as an Archimedean point outside of everything given, is clearly quite foreign to the classical way of considering human action.

---

22 L. Strauss, *Thoughts on Machiavelli* (Glencoe: The Free Press, 1959), p. 298.

Such a wide difference of doctrine illustrates the difficulty of distinguishing the "quaestio facti" (did the classical philosophers think thus?) from the "quaestio juris" (were they right to think so?). The fundamental nature of the presuppositions involved for a contemporary commentator in examining the "quaestio juris" may so dominate the mind as to make difficult the objective scrutiny of the "quaestio facti." On the one hand, it has been possible for modern scholars to believe that there could be no cogent reasons for turning one's back on so obvious a good as the conquest of nature, and so to read the Greek political philosophers as if they were a preparation for the greater wisdom of the age of progress. On the other hand, it would be possible to say that Strauss is reading into these references from the Greeks a clarity about their rejection of the conquest of nature which is not present in their writings, and that he does so because of his concern to show a consistent alternative to the modern conception of the science of nature. The justification of Strauss's assertion of fact cannot be expected to be found solely in the scrutiny of any particular text.[23]

Having said this, however, I must emphasize that the question of fact requires more elucidation than Strauss gives it. It requires, for instance, some discussion of Greek mathematics. Strauss's thesis at this point is based on the assumption that what the Greeks turned away from was not inventions in general but the use of science for such inventions. "Such use of science is excluded by the nature of science as a theoretical pursuit."[24] But to understand such a statement as this it would be necessary to understand Greek geometry and what those Greeks who were philosophers thought geometry was and also what place geometry played in Greek religious practice before Aristotle. As a student of religion, it is quite clear to me that geometry had for the educated Greeks an essentially religious

---

23 The question here must be related to Strauss's doctrine of hermeneutics: read the works of any great philosopher with the certainty that one is not in a position to understand better what the philosopher is saying than he is himself. "I do not know of any historian who grasped fully a fundamental presupposition of a great thinker which the great thinker did not grasp." See *What Is Political Philosophy?*, p. 228. This remark of Strauss is made in answer to Professor G. H. Sabine's criticism of him which includes the following statement: "There are presuppositions implicit in.... 'the climate of opinion' of any age that no contemporary ever fully grasps."

24 Strauss, *Thoughts on Machiavelli*, p. 299.

significance, but what that significance was I have not sufficient knowledge of either their geometry or their religion to understand as yet. One could wish, therefore, that even if Strauss did not include a discussion of Greek geometry in his text that he had included references to scholarly writings which would illuminate what he says about Greek geometry and its relation to their philosophy and religion.[25]

To turn directly to the "quaestio juris," I will state in simple language what Kojève and Strauss seem to be saying about society and technology. According to Kojève, unlimited technological progress is not worth pursuing for its own sake. It should serve human interests beyond itself because it should be pursued until there is no more work for men to do. According to Strauss, modern men are committed to unlimited technological progress. Here it may be mentioned that in the age of space investigation Strauss's account of modern men seems more accurate than Kojève's. It would appear to me that technological progress is now being pursued not first and foremost to free all men from work and disease, but for the investigation and conquest of the infinite spaces around us. The vastness of such a task suggests that modern society is committed to unlimited technological progress for its own sake.

My difficulty in comprehending Strauss's position lies not then in giving some meaning to the idea that the dominant leaders of our society are committed to unlimited technological progress, but rather in understanding what it meant to the classical political philosophers not to be so committed, and even more in understanding what it would mean not to be so committed in the contemporary world. Strauss makes three other points germane to this subject.

(1)   "The classics were for almost all practical purposes what now are called conservatives. In contradistinction to many present day conservatives, however, they knew that one cannot be distrustful of political or social change without being distrustful of technological change. Therefore they did not favour the encouragement of inventions, except perhaps in tyrannies, *i.e.*, in regimes the change of which is manifestly desirable. They demanded the strict moral-political supervision of inventions; the good

---

25   For an essay on this matter which reaches a similar conclusion see Simone Weil, *Intuitions Pré-Chrétiennes, A propos de la doctrine pythagoricienne* (Paris: La Colombe, 1951).

and wise city will determine which inventions are to be made use of and which are to be suppressed."[26] To ask the question, by what criteria the rulers of the good and wise city were to make these determinations, or did in fact make these determinations, would, I presume, draw from Strauss the reply: by that virtue and piety which are described in the leading classical books on moral and political philosophy. The issue then returns to the completeness, adequacy and concreteness of that teaching. Strauss's position would be easier to understand if he would explicate the classical teaching on this matter.

(2)   The second point can best be made by quoting from Strauss directly:

"The opinion that there occur periodic cataclysms in fact took care of any apprehension regarding an excessive development of technology or regarding the danger that man's inventions might become his masters and his destroyers. Viewed in this light, the natural cataclysms appear as a manifestation of the beneficence of nature. Machiavelli himself expresses this opinion of the natural cataclysms which has been rendered incredible by the experience of the last centuries. It would seem that the notion of the beneficence of nature or of the primacy of the Good must be restored by being rethought through a return to the fundamental experience from which it is derived."[27]

(3)   Strauss's third point concerns the proper role of the philosopher in a society the majority and dominating minority of which believe in unlimited technological progress. He states by implication that the philosopher who recognizes that society is committed to such an unwise pursuit must not fall into the temptation of escaping into anti-social dreams.[28] In a sentence at the end of his French essay Strauss writes of Kojève and himself:

"Mais nous y avons toujours été attentif, car nous nous détournons tous deux, en apparence, de l'Être pour nous tourner vers la tyrannie, parce que nous avons vu que ceux qui manquent

---

26   *Thoughts on Machiavelli*, p. 298.

27   *Ibid.*, p. 299.

28   This point is made in a review by Strauss of a book by Yves Simon as found in *What Is Political Philosophy?*, p. 306 *et seq.*

de courage pour braver les conséquences de la tyrannie, qui,
par conséquent 'et humiliter serviebant et superbe domina-
bantur,' étaient forcés de s'évader tout autant des conséquences
de l'Être, précisément parce qu'ils ne faisaient rien d'autre que
parler de l'Être." [29]

These three comments of Strauss help to make clear what he
considers the classical political philosophers' views of the proper
relation of society and technology. The difference between them and
the views advocated by Kojève is of course too deep for it to be possible
to argue their respective merits in the course of an article. Nevertheless
there is one argument on the modern side which the interests of
charity require should be presented. It is the following: no writing
about technological progress and the rightness of imposing limits
upon it should avoid expressing the fact that the poor, the diseased,
the hungry, and the tired can hardly be expected to contemplate
any such limitation with the equanimity of the philosopher. Strauss
is clearly aware of this fact. [30] One could wish however that he had
drawn out the implications of it in the present controversy. It is not
by accident that as representative and perceptive a modern political
philosopher as Feuerbach should have written that "compassion
is before thought." The plea for the superiority of classical political
science over the modern assumptions must come to terms with the
implications of this phrase in full explicitness. As the assertion that
charity is more important than thought is obviously of Biblical origin,
his point leads directly to my second area of commentary.

(B)   The second comment I wish to make concerns the interpretation which Kojève and Strauss place on the history of western thought
and in particular their interpretation of the relationship between the

29 *Les Essais,* p. 344. What Strauss and Kojève have been attentive to is their
different hypotheses about philosophy. Strauss's Latin tag in this sentence is
an adaptation of Livy's remark about the masses. It also indicts individual
men such as Martin Heidegger. I have not translated this passage because I
wish strongly that Strauss would include it in any reissue of his English essay.

30 L. Strauss, "On a New Interpretation of Plato's Political Philosophy," *Social
Research,* June 1946.

history of philosophy and Biblical religion.[31] The issue is clearly present in the controversy. At the beginning of his essay Strauss summarizes the gist of the only valuable criticisms of his writing on *Hiero* (namely those of Voegelin and Kojève): "Is the attempt to restore classical social science not utopian since it implies that the classical orientation has not been made obsolete by the triumph of the Biblical orientation?"[32] The understanding of the controversy therefore requires an understanding of what they both mean by the Biblical orientation.

To repeat, Hegel's moral-political teaching is definitive for Kojève and it claims to be a synthesis of Greek and Biblical teaching. The Greek morality of honour is the morality of the master, and it meets its antithesis in the morality of the slave basic to Biblical theism. The synthesis of these is found in the absolute morality of *The Phenomenology of Spirit*, in which the dialectic of master and slave, and the argument which follows from it, achieve that synthesis. The centre of Kojève's argument is the superiority of Hegelian social science over the classics, because the former included in itself the understanding of the morality of the slave which is central to Biblical theism and which is a higher moral standpoint than the morality of the master. Hegel's political doctrine was a rationalizing, *i.e.*, a secularizing of Biblical theism, and its synthesising with the Greek morality of honour.[33]

Strauss on the other hand interprets Hegel's moral-political teaching as founded on Machiavellian or Hobbesian teaching. He maintains that the teaching about master and slave is directly based on Hobbes' doctrine of the state of nature. That is, Hegel as much as Hobbes constructs his political doctrine on the assumption "that man as man is thinkable as a being that lacks awareness of sacred restraints or as a

---

31 When I use the word Biblical in this essay I use it as a common term referring to the religion of the Old and New Testaments. Any possible fundamental differences between Judaism and Christianity will neither be discussed nor assumed.

32 *What Is Political Philosophy?*, p. 96.

33 These few words must be read along with *The Phenomenology of Spirit* and Kojève's commentary on it. Hegel's work must also be read in the light of his early writings about Christianity.

being that is guided by nothing but a desire for recognition."[34] But the Hobbesian doctrine of the state of nature cannot be reconciled with the conception of nature common to the classical political philosophers, who asserted the beneficence of nature or the primacy of the Good, nor indeed with the piety of Greek religion.

If this be so (and I think it is) then the Hegelian account of western philosophy and its relation to Biblical religion is clearly not true, because it is not possible to synthesize Socratic and Hobbesian politics. But the question still remains whether the Machiavellian and Hobbesian politics are at least in part a result of the Biblical orientation of western society.

For an answer to what Strauss thinks of this question it would be necessary to comment in detail on his books on Hobbes and Machiavelli, and this of course cannot be done here.[35] Nevertheless it is possible to say something about the question within the limits of what Strauss says in this controversy. Strauss maintains that the political teaching of Hegel-Kojève is based on the substitution of universal recognition for the classical goal of moral virtue as the most satisfying object of human striving. Indeed he asserts that the desire among modern political philosophers to actualize the best social order was influential in the lowering of their vision of that highest goal, so that this actualising would not be impossible. This has led in modern political theory to the emancipation of the passions and an unfounded optimism about the effect of the movements instigated by the passions. Strauss writes: "Syntheses effect miracles. Kojève's or Hegel's synthesis of classical and Biblical morality effects the miracle of producing an amazingly lax morality out of two moralities both of which made very strict demands on self-restraint. Neither Biblical nor classical morality encourages us to try, solely for the sake of our preferment or our glory, to oust from their positions men who do the required work as well as we could. Neither Biblical nor

---

34 *What Is Political Philosophy?*, p. 111.

35 Strauss's book on Hobbes was published in 1936 and his book on Machiavelli in 1959. It is a fair inference therefore that his maturer doctrine on this matter will be found in the later book. It is also worth stating that Strauss was led to the study of Hobbes from Spinoza, and particularly from Spinoza's methods of Biblical criticism.

classical morality encourages all statesmen to try to extend their authority over all men in order to achieve universal recognition."[36]

This makes clear that Strauss does not identify the Biblical orientation with the emancipation of the passions, so characteristic of modern moral philosophy since Spinoza. Yet the rejection of the Hegelian account of the relation between modern philosophy and Biblical religion still leaves one with the question of what that relation has been. This question cannot be avoided by a thinker such as Strauss, who is attempting to restore classical social science. The impossibility of that avoidance can be seen in one platitudinous generalization: one difference between all European philosophy up to the twentieth century and classical philosophy is that the former was written by men who lived in a society permeated with Biblical religion. An historian has recently written, "By the middle of the thirteenth century, a considerable group of active minds...were coming to think of the cosmos as a vast reservoir of the energies to be tapped and used according to human intentions."[37] If this statement is true, and if (as I have already quoted from Strauss) "Modern man as little as pre-modern man can escape imitating nature as he understands nature," then clearly the question arises as to the connection between the religion of western Europe and the dynamic civilization which first arose there, the spread of which has been so rapid in our century. This is the civilization which in the opinion of both Strauss and Kojève tends towards the universal and homogeneous state. Comment A about technology is therefore intimately related to comment B about the Biblical orientation.

The question as to what Strauss understands by the relation of philosophy to Biblical religion must be prefaced by one obvious but necessary caveat. Just as philosophy has always been problematic to itself, so equally Biblical religion is not an easily definable entity, either for those who conduct their life of worship within its terms or for those outside it. We cannot write of Biblical religion as if it were something so

---

36 *What Is Political Philosophy?*, p. 111.

37 L. White, *Medieval Technology and Social Change* (Oxford: Oxford University Press, 1962) pp. 133–134.

obviously given that such concepts as "theism," "creation," "revelation," etc., have some common meaning from which further debate can proceed. Indeed both Jewish and Christian philosophers on the one hand and secular philosophers on the other have often increased the difficulty of understanding the past by writing as if the entity "Biblical religion" was a clear and distinct idea. To use religious rather than philosophical language: is Christianity a worldly or an unworldly religion? [38] Within the Christian tradition, believers have obviously interpreted their religion in both these ways and in varying compromises. Is Christianity fundamentally oriented to history or to eternity? [39] Moreover there has always been a wide disparity of theological opinion in the Christian tradition about the relation of the Biblical doctrine of the Fall to the classical doctrine of the beneficence of nature. Or again, is or is not Christianity's moral position rightly described as that of the slave? I raise these obvious points for the following reasons: in asking what Strauss thinks about the role of Biblical religion in western history, I am also asking what he considers Biblical religion to be. The effort to understand Biblical religion is as much a philosophical task as to understand its relation to the pursuit of wisdom. For instance, what has been said above in criticism of Hegel is not simply that he fails to synthesize Greek and Biblical morality, but that he holds an incomplete and one-sided account of Biblical theism itself, and that certain errors in his political philosophy stem from that misinterpretation.

To turn to the question itself, I can only state that Strauss's writing shows a remarkable reticence whenever he writes of Biblical religion, and particularly about the authority that the Bible should have for western men. He writes with sympathy of Maimonides (a philosopher whom he

---

38 I speak here of Christianity, rather than of Judaism, because it is less impertinent to raise dilemmas about one's own allegiance than about other peoples'.

39 There is in our era in both Catholic and Protestant theological circles a strong emphasis on Christianity as an historical or worldly religion. The way this is expressed seems often little more than an attempt to justify both Christianity and our present dynamic industrial civilization. This combined justification may be wise but it is certainly not easily reconciled with what one reads in the Greek Fathers or even for that matter with the work of the father of western theology. See St. Augustine, *De Civitate Dei*, Book 22, para. 24.

considers more profound than Spinoza). Also in a lecture delivered in Jerusalem he has stated: "Nowhere else has the longing for justice and the just city filled the purest hearts and the loftiest souls with such zeal as on this sacred soil." [40] But beyond such general statements, I find it impossible to know whether he thinks there is in the Bible an authority of revelation which has a claim over the philosopher as much as over other men. Nor can I tell, on the other hand, whether he thinks that such an assumption of authority must be for obvious reasons inimical to the claims of philosophy. Indeed there are in his writings occasional passages where he shows contempt for certain forms of Biblical religion. [41]

Reticence on such an important subject by as subtle and definite a writer as Strauss must be taken as indicative of an implied position. His work as a whole has been concerned with a careful rediscovery of the teachings of the past so that political philosophers will not simply accept the contemporary presuppositions. About other aspects of the tradition he speaks unambiguously.

It is perhaps not unwise to hazard an explanation for this lacuna. This does not entail any such folly as the seeking out of some irrational explanation based on the assumption of a fuller knowledge of Strauss's psyche than he has himself. The twentieth century has already had its bellyful of such impudent psychologizing. Rather I mean to state certain propositions which Strauss affirms to be true and draw from them a conclusion which would explain his reticence. According to Strauss, one of the essential truths of classical political philosophy is that all men are not capable of becoming philosophers. Indeed nothing is more essential for the proper ordering of society than that this distinction in nature should be understood and preserved. Also he agrees that the health of the social order can be based only on piety and virtue, and not on socially useful passions, as is presumed by most modern political philosophers. Taken together these two statements imply that in any passably good social order there will be among the majority some practice of religion. Indeed he makes clear that one of the essential points on which modern political philosophy broke

---

40 *What Is Political Philosophy?*, p. 9.

41 *Ibid.*, p. 23.

with the classical teaching was in its belief: "Philosophy is to fulfill the function of both philosophy and religion."[42] It is also the case that in the western world what remnants of sacred restraints still linger in the minds of men are most often connected with the practice of the two religions, Judaism and Christianity, which alone are indigenous to the western world. Therefore, even if Strauss should in fact think that the Biblical categories have been in part responsible for a false and therefore dangerous conception of nature among modern philosophers, he would not necessarily think it wise to speak openly or forcibly about the matter. He is a philosopher, and not one of those who consider it their function and their joy to "enlighten" the majority by undermining their trust in the main religious practice which is open to them. Moreover, no writer has so emphasized the various reasons why philosophers have maintained and will continue to maintain a distinction between their exoteric and their esoteric teaching.

Strauss's reticence about Biblical religion puts him at a disadvantage in his argument with Kojève. Kojève's account of the place of Biblical religion in western history is quite explicit. Strauss is able to criticize (and in my opinion with success) that account of western history. Because he does not speak clearly about Biblical religion, however, he is not able to state, except by implication, his own account of that history. Indeed, I would go farther. Since Strauss is attempting the remarkable and prodigious task of restoring classical social science, how can he maintain his reticence at this point?

---

42 *Thoughts on Machiavelli*, p. 297.

The University Curriculum

THE CURRICULUM IS THE ESSENCE of any university. It consists in what students formally study at all stages from the undergraduate to the research professor. It determines the character of the university far more than any structure of government, methods of teaching, or social organisation. Indeed, these latter are largely shaped by what is studied and why it is studied. The curriculum is itself chiefly determined by what the dominant classes of the society consider important to be known.

Members of the dominant classes make the decisions which embody the chief purposes of any society, but their very dominance is dependent on their service of those purposes. The primary purpose in Canadian society is to keep technology dynamic within the context of the continental state capitalist structure. By technology I mean "the totality of methods rationally arrived at and having absolute efficiency (for a given stage of development) in every field of human activity."[1] The dynamism of technology has gradually become the dominant purpose in western civilization because the most influential men in that civilization have believed for the last centuries that the mastery of chance was the chief means of improving the race. It is difficult to estimate how much this quest for mastery is still believed to serve the hope of men's perfecting, or how much it is now an autonomous quest. Be that as it may, one finds agreement between corporation executive and union member, farmer and suburbanite, cautious and radical politician, university

---

1 See J. Ellul. *The Technological Society,* London 1965, p. xxxiii.

administrator and civil servant, in that they all effectively subscribe to society's faith in mastery.

The phrase "state capitalist structure" is necessary because although the pursuit of a dynamic technology is now a world wide religion, it is carried on within a different structure in North America than in Asia or eastern Europe. That system is based primarily on co-ordination of power between private and public corporations and can best be called "state" or "corporation" capitalism. By "continental" I mean that the Canadian structure can only be understood as a satellite of the broader imperial system.[2]

The chief purpose of the curriculum in Canadian universities is, then, to facilitate the production of personnel necessary to that type of society. Because there is such agreement about the chief goal of society, there is a vast consensus about the principles of the curriculum. Debates take place about the government of the university, about humane existence within it, etc., etc., but not about what it concerns a human being to know. So monolithic is the agreement of society about ends, so pervasive the ideology of liberalism which expresses that agreement, that the question about knowing cannot be raised seriously.[3] When a student first arrives at university, the curriculum may appear a set of arbitrary and incoherent details. This is so only at the surface. In fact, it can be understood in terms of the powers and purposes of our society.

The foregoing generalizations would, of course, require careful application in the case of any particular university. For example, the phrase "dominant classes" cannot be easily specified. It is clear that the governments of the provinces (*i.e.*, the professional politicians and the educational civil servants) have an increasing influence over the

2 Because the U.S. is an imperial power and Canada is not yet entirely integrated into all the purposes of that empire, the garrison-state qualities in the American educational system are not yet as advanced in Canada. Not only do we have no draft, but fewer military needs have to be served by the universities, whether in research or teaching.

3 I mean by liberalism a set of beliefs which proceed from the central assumption that man's essence is his freedom and therefore that what chiefly concerns man in this life is to shape the world as we want it.

curriculum as compared with that held by the representatives of the private corporations on the Boards of Governors. But the question cannot be dropped at this point, because that leaves unclear the relation of professional politicians and civil servants to private corporations in a state capitalist society. And the situation varies in various parts of Canada. The relation between provincial governments and private corporations varies greatly in Alberta, Ontario, and Nova Scotia. (Because of the different traditions in Quebec, this article is not concerned with French-Canadian curricula.) Boards of Governors which include the rulers of Brazilian Traction will be more formidable exercisers of power than boards composed of local civil servants and worthies. Or, another illustration of the need to specify: the old-fashioned literary education (classics, modern history and literatures, the history of philosophy) still maintains some prestige at the University of Toronto. The weight of tradition carries on in an established university for several generations, with the result that the curriculum may reflect the ideas of a class which is no longer dominant outside its walls. Only slowly and often almost imperceptibly do curricula respond to the powers and purposes of the society.

The chief job of the universities within the technological societies is the cultivation of those sciences which issue in the mastery of human and non-human nature. It is clearer to describe the sciences according to their part within this purpose of mastery than by the older rubric of "natural" and "social" which described the sciences according to whether their subject matter was concerned with the human or the non-human. For example, biochemistry may be said to be concerned with the chemistry of life and therefore be included under natural science. It is in fact concerned more and more with humanity — indeed with the very roots of humanness. What could be more a "social science" than one which opens the possibility that sun and man will no longer generate man? The fact that the categories natural and social are no longer adequate divisions can be seen in the necessity of a new category, "the health sciences," under which much of the work of biochemistry must fall, and which is a mediating category between natural and social. Also, psychology (traditionally a social science) is now largely taken up with questions which once would have fallen clearly within physiology. At the same time, the understanding of the physiology of the human brain must mean, in its direct relation to mastery, that modern psychology is as much a social science as it ever was. Indeed, the modern unity of the sciences is realized around the ideal of mastery. And this is not negated by the necessary proliferation of new specialisms, "biophysics," "chemical physics," etc., etc., etc. The proliferation of division is accompanied by the equally necessary proliferation of the interdisciplinary unit which holds the studies together around the varying means of mastery.

It would, of course, be absurd to deny that the pure desire to know is present in many modern scientists. In my experience, such a desire exists in the community of natural scientists more than in any other group in our society. Also, I would assert on principle that such a desire belongs to man as man. What I am saying is that in North American science the motive of wonder becomes ever more subsidiary to the motive of power, and that those scientists still dominated by wonder have a more difficult time resisting the forces of power which press in upon them from without their community. For example, the recently established Science Council of Canada is surely intended to integrate the scientific community into the pragmatic purposes of the private and public corporations. It is this growing victory of power over wonder which is the basis of the proposition that the modern sciences can best be understood as a unity around the idea of mastery.

The job of producing personnel who know something about the control of human and non-human nature requires teaching and research at many levels from the highly theoretical to the immediately utilitarian. Those who study mathematics in its most formal mode sometimes assert that they have little interest in the relation of their studies to its uses. It is hard to deny, however, that the privileged place of mathematics in modern curricula is related to the fact that algebra (employ if you will a more modern name) has uses in the technological society. Von Neumann and Wiener were not exactly neglected by the powerful. There are some physicists whose cosmological investigations seem far from the practical and whose intention is clearly the pure desire to know. Nevertheless, the interdependence of modern physics and the technological society is evident to common sense. Oppenheimer, who spoke often and eloquently to his admiring scientific colleagues about the beauty of science as a contemplative discipline, still made clear that putting nature to the question is of the essence of modern science when he said the terrible words: "If an experiment is sweet, one must go ahead with it." Moreover, the theoretical physicists are increasingly out-numbered in their community by research teams whose work is more immediately applicable to the conquest of nature.

Physics and chemistry and biology departments must aid in the production of an enormous range of personnel reaching out from their own discipline towards other faculties of the university, particularly medicine and engineering. They must train experts to advance knowledge by research in their increasingly complex subjects. They must produce enough specialists to maintain the tradition in the high schools. They must teach as much of their subject to engineers, doctors, subsidiary therapists, etc., as is necessary for the practice of these professions. They must introduce their subject in general courses for the undergraduate community as a whole. Indeed, on the one side the basic science departments must in their membership narrow down to increasingly detailed concentration on one facet of their complex science, so that the university proliferates with new divisions which have in some sense fragmented off from the original unity of the subject; on the other hand, they broaden out to wider and wider reaches of the population who need some understanding of basic science if they are to exist as personnel in our society. [4] Yet, both the narrowing of subject matter and the broadening of clientele are results of the same principle: the purpose of education is to gain knowledge which issues in the mastery of human and non-human nature.

Within the last hundred years, it has become increasingly clear that the technological society requires not only the control of non-human nature, but equally the control of human nature. This is the chief cause of the development of the modern "value-free" social sciences. Here again, the distinction must be made in the work of departments such

---

4 Paul Goodman has often made the point that much of the academic training required of the young is not necessary for the practice by which they will earn their living. The degree to which Goodman's point is true in opposition to my account could only be decided by a detailed description of various occupations and the degree to which their practice is dependent on knowledge of the various sciences.

as sociology and psychology between the function of expanding that knowledge which gives control, and the need to produce innumerable personnel who apply current knowledge. A society in which there are more and more people living in closer and closer proximity will need enormous numbers of regulators to oil the works through their knowledge of intelligence testing, social structures, Oedipal fixations, deviant behaviour, learning theory, etc. The old adage about the need for more science to meet the problems that science has created will be illustrated in the proliferation of these techniques. The same applies to the more traditional science of economics. The household is now imperial and its management requires innumerable accountants, whether they call themselves econometrists or doctors of business administration.

For the social scientists to play their controlling role required that they should come to interpret their sciences as "value-free." This clarification has been carried out particularly by sociologists, and, indeed, it is inevitable from its very subject matter that this science should be magisterial among the social sciences. The use of the term, "value" and the distinguishing of judgements about values from judgements about facts enables the social scientist to believe that his account of reality is objective, while all previous accounts (which were not based on this distinction) were vitiated by their confusion of normative with factual statements. It is not appropriate in this writing to describe the history of the idea of "value-free" social science as it came to be in the European tradition, particularly under the influence of Kant and Nietzsche, and was so elegantly formulated by Weber. Nor is it necessary to describe how it has been reformulated in liberal Protestant terms by such men as Parsons and Lasswell, to suit the American scene.[5] What is important to understand is that the quantification-oriented behavioural sciences which have arisen from this methodological history are wonderfully appropriate for serving the tasks of control necessary to a technological society. Indeed, where the fact-value distinction was originally formulated by Weber as a means whereby the academy could hold itself free from the pressures of the powerful, it has quickly become in North

5   To describe this intellectual movement whereby the Y.M.C.A. and Nietzsche were brought together would require an art beyond me — that of comedy.

America a means whereby the university can make itself socially useful. Social sciences so defined are well adapted to serve the purposes of the ruling private and public corporations.

Indeed, the distinction between judgements of fact and judgements of value has been thought to be favourable to a pluralist society. The common or objective world would be that of facts known scientifically, leaving men free to choose their values for themselves. However, this distinction has worked in exactly the opposite direction towards the monism of technological values. From the assumption that the scientific method is not concerned with judgements of value, it is but a short step to asserting that reason cannot tell us anything about good and bad, and that, therefore, judgements of value are subjective preferences based on our particular emotional makeup. But the very idea that good and bad are subjective preferences removes one possible brake from the triumphant chariot of technology. The rhetoric of pluralism simply legitimizes the monistic fact.

The "value-free" social sciences not only provide the means of control, but also provide a large percentage of the preachers who proclaim the dogmas which legitimize modern liberalism within the university. At first sight, it might be thought that practitioners of "value-free" science would not make good preachers. In looking more closely, however, it will be seen that the fact-value distinction is not self-evident, as is often claimed. It assumes a particular account of moral judgement, and a particular account of objectivity. To use the language of value about moral judgement is to assume that what man is doing when he is moral is choosing in his freedom to make the world according to his own values which are not derived from knowledge of the cosmos. To confine the language of objectivity to what is open to quantifiable experiment is to limit purpose to our own subjectivity. As these metaphysical roots of the fact-value distinction are not often evident to those who affirm the method, they are generally inculcated in what can be best described as a religious way; that is, as doctrine beyond question.

I do not mean to imply any insincerity on the part of those teachers who preach this doctrine. Throughout history the best preachers have often been those who thought they were talking about universally self-evident facts. Nor should it be implied that our multiversities ought not to fulfil this legitimizing role. Class liberalism is the ideological cement for a technological society of our type. Its sermons have to be preached to the young, and the multiversities are the appropriate place. That the clever have to put up with this as a substitute for the cultivation of the intellect is a price they must pay to the interests of the majority who are in need of some public religion. And the fact-value distinction is the most sacred doctrine of our public religion.

To turn to "the humanities": their place in the curriculum is difficult to state because many different accounts of their purposes are given by their practitioners, and some of these accounts seem remote and disconnected from the age in which we live. Indeed, in thinking about the humanities, the intellectual uncertainty faced by those who grow up in this era becomes most apparent. The root form of this uncertainty lies in the question as to what of importance can be known other than that which is given in those sciences which proceed from quantifying and experimental methods. Because of the difficulty of this question, a short word about the past is here required.

In the antique world, it was assumed (and that assumption was most fully articulated by Socrates) that the purpose of education was the search through free insight for what constituted the best life for men in their cities. Such education was reserved for the few, because free insight was possible only with leisure, and the ancient world could only achieve leisure for some. Such education was concerned not only with human concerns but with the non-human, because it was thought that man could understand what was best for himself and the species only if he understood the cosmos as a whole. All the arts and sciences (and the word science meant any systematic body of theoretical knowledge) depended on the highest science, "philosophia," which was thought to give one knowledge of the whole or at least openness to the whole, in the sense of openness to the most important questions. When this scheme of education was taken over by civilizations dominated by some form of Semitic religion, it was not altogether changed. The belief that there had been given divine revelation which told men what was most important for them to know meant that the free insight of philosophy became subservient to that revelation. Yet, such civilizations, whether consistently or not, often maintained philosophy in the curricula as the way that educated men could make that revelation intelligible to themselves and others. The liberal arts continued to be seen as a preparation for that philosophy. Whatever the tension between philosophy and revelation in western society, education under their combined rubric was directed to knowledge of man's highest end.

It is not possible in this article to state how or why the quantifying and experimental methods have become dominant in the sciences, or how this development is related to the freeing of the particular sciences from the magistery of philosophy or to relate both these to the falling away of any belief in revelation. What is important in the present connection is to insist that these occurrences have put in question (a) whether there is any knowledge other than that reached by quantifying and experimental methods, (b) whether, as such methods cannot provide knowledge of the proper purposes of human life, the very idea of there being better and worse purposes has any sense to it, (c) whether, indeed, purpose is not merely what we will in power from the midst of chaos. The effect of these questionings on the humanities could not but be enormous.

In the last hundred years in Europe, a series of justifications of humane study arose in the light of the crisis produced by the age of progress. Each of these passing justifications made certain particular studies dominant for their particular hour. For example, Dilthey's distinction between "Naturwissenschaft" and "Geisteswissenschaft" led to the enormous concentration on the study of history as that which would fulfil the role which once had been played by the traditional philosophy and theology. By studying history men could understand the alternatives of the past, see where they were and be enlightened to choose where they were going. The humanities became the sciences of the human spirit which culminated in that new subject, the philosophy of history. This position was in turn destroyed by Nietzsche when he showed that history could not, any more than God, provide men with a horizon within which to live. In terms of this critique, Weber taught that a humane and scientific sociology could fulfil the magisterial role. Over the century these various justifications have had their necessary moment, but as they have succeeded each other, the humanities have become a smaller and smaller island in a rising lake. The drowning lake was the ever more clearly formulated assumption that all the important questions can be solved by technological means.

The North American situation both followed and differed from Europe. In the older world, institutions and traditions which incarnated the old philosophical education could not be quickly swept away. Even such thinkers as Nietzsche and Russell who were in different ways criticizing out of existence the old moral and religious basis, still in some deep sense believed and depended on the contemplative tradition the remnants of which they were demolishing. As we can see from our present situation, the Calvinist pioneers were building in English-speaking North America a society which would be more completely and quickly dominated by technology than any other. Yet, at the same time, the absence of philosophy in North America meant an absence of the extreme forms of nihilism. North American society was till recently both more innocent and more barren than Europe.

In so far as there was a public or popular justification of the humanities in North America, it could be heard in those endless convocation addresses under which men suffered from 1900–1950. "The humanities will teach us to choose whether we will use our immense technical power for good or ill" etc., etc. This justification was allied with the general progressive hope that as technology advanced, men in their leisure would share in the riches of culture. (What exactly these riches were was rarely specified.) It was also sometimes allied to the democratic belief that all men should share in their governing and that for them to do so would require an education which would transcend the simply technical. Indeed, in many liberal minds widespread university education was seen as fulfilling the role which had been played by revelation in the once dominant Calvinist Protestantism. When revelation no longer appeared to be a fact to the powerful classes, the hope appeared among some that the humanities in the universities would teach men the best purposes. This popular hope could never be realized for the following reason: those who knew the humanities professionally were aware of what was going on in Europe. The best of them knew that social thought was methodologically dominated by "historicism" and "the fact-value distinction." Historicism was the belief that the values of any culture were relative to the absolute presuppositions of that culture which were themselves historically determined, and that therefore men could not in their reasoning transcend their own epoch. The fact-value distinction led generally to the conclusion that there was no rational way of knowing that one way of life was nobler than another. Those who studied the humanities were led to a great uncertainty about what constituted the good life, and whether this was a real question at all. The public hope that the humanities would fulfil a positive moral role was, therefore, vitiated by the fact that the best professionals of these disciplines did not see their activity in this way. Indeed, it is a fact of North American history that the spread of professional teaching in the humanities has been a means whereby the scepticism of Europe has penetrated the more innocent traditions of North America.

The professional practitioners of the humanities have justified their studies quite differently from the popular rhetoric. They have increasingly said that the humanities are non-evaluative sciences. The cruder form of this justification has led those disciplines to become highly research-oriented, so that they could cover themselves with the mantle of science and Protestant busyness. An enormous amount of energy and money has been channelled into research projects. In English literature there are many great factories pouring out editions, commentaries, and lives on all but the miniscule figures of our literature. (The equivalent of the expensive equipment of the scientists are the rare manuscript libraries.) If one has a steady nerve, it is useful to contemplate how much is written about Beowulf in one year in North America. One can look at the Shakespeare industry with perhaps less sense of absurdity; but when it comes to figures such as Horace Walpole having their own factory, one must beware vertigo. The difficulty in this research orientation is that whereas research in the progressive sciences produces discoveries which the public see as useful, this is not so in the humanities. The historian may claim that all the careful work that goes into small areas can be justified as useful in that it makes up the totality of a mosaic from which those who are educated in the discipline may better know the past and so make more prudent judgements in the present. He may even claim that the formal discipline of Namerian history is itself a good training for potential rulers. Both these justifications may be true, although the proof of this would require a discussion of the place of historical judgement in the training of rulers. What is, however, paradoxical at the practical level about the vast apparatus of modern historical scholarship is that it exists in an age when it is increasingly believed that the race has little to learn from human existence from before the age of progress. We may be grateful that this contempt for the past has not yet penetrated the ranks of academic historians, who therefore remain the chief brake on the simply modern in our multiversities. Nevertheless, the role of brake is a very minor one compared with the pre-progressive role of the humanities.

The justification of the humanities as sciences takes a deeper form than simply concentration on research. Scholars turn more and more to the practice of non-evaluative analysis. From literary criticism this can be well illustrated in the work of Northrop Frye, in which the study of literature becomes a classificatory science with the claims to objectivity and progress which go with such a science. In a rather similar way, the study which still uses the name of "philosophy" has made itself into a particular science, with its own particular rigors, concerned with the analysis of language, methods, and thought. Indeed, in so far as philosophy moves beyond this non-evaluative analysis it is concerned with what has been done under its name in the past, and this history is more and more seen as a series of inadequate logical formulations which can be corrected in the light of advances in analysis.

Non-evaluative analysts see their activity as essentially self-justifying. They are moved by the pure desire to know; for example, the sheer joy in mastery over such a diverse field as literature. This gives the humanities freedom from the crude pressures of society, as can be seen by comparison with the popular justification of the humanities which has been described earlier. In the popular account, the natural scientists were supposed to know so that they could provide material techniques; the teachers of humanities were supposed to know so that they could be purveyors of values. In both cases, the pure desire to know was considered subsidiary to some public end outside the subject itself. Non-evaluative analysis has exalted cognitive power and has brought back into the disciplines rigor which was often lacking when professors of literature and philosophy were surrogate preachers. Although it must be granted that non-evaluative analysts can, at their worst, fall into the snobbishness of an impractical mandarinism, they have saved the humanities from an empty antiquarianism. They are interested in the understanding and mastery of literature as a living activity and in the practice of linguistic clarity as important in the present.

Nevertheless, the consequences of this approach must be insisted upon. Non-evaluative analysis cuts men off from openness to certain questions. Let us imagine a student who is studying the works of Tolstoi and de Sade under the guidance of an intelligent and sensitive practitioner of non-evaluative analysis. He can be taught to understand much about their writings, what is being said and how it is being said and the dependence of these on very complex traditions. As in some sense both authors are writing about the proper place of sexuality in human existence, the student can be taught to anatomize the similarities and differences in what the authors say about sexual "values." From such study the student will learn what two remarkable men have thought about the place of sexuality in human life. Yet, because the study is a non-evaluative science, what would seem to be the most important question cannot be raised within the study: that is, whether de Sade or Tolstoi is nearer the truth about the proper place of sexuality. In the same way in "philosophy" the study of ethics tells one much of how language is used and can be used consistently in ethical discourse. But it no longer claims to be concerned with what are the highest possibilities for men. Such studies are impotent to lead to what was once considered (perhaps and perhaps not naively) the crucial judgement about "values" — whether they are good or evil. Their scholars have gained their unassailable status of mastery and self-justification by surrendering their power to speak about questions of immediate and ultimate meaning — indeed generally by asserting that such questions only arise through confusion of mind. Such a position provides immunity within the academic fortress, but it can still be asked whether the impotence of mind towards meaning is man's necessary condition.

Be that as it may, the central role of the humanities will be increasingly as handmaiden to the performing arts. To repeat, the dominant ethos in the society is provided by an autonomous technology. But the space programme, necessary imperial wars, and the struggle for recognition in the interlocking corporations can provide purpose only for a small minority. Purpose for the majority will be found in the subsidiary ethos of the fun culture. It will meet the needs of those who live in affluence but are removed from any directing of the society. One is tempted to state that the North American motto is: "the orgasm at home and napalm abroad," but in the nervous mobile society, people have only so much capacity for orgasm, and the flickering messages of the performing arts will fill the interstices. They provide the entertainment and release which technological society requires. The public purpose of art will not be to lead men to the meaning of things, but to titivate, cajole, and shock them into fitting into a world in which the question of meaning is not relevant. The humanities in the universities will become handmaidens in this task. This will not mean that they will be weak in numbers or prestige or resources because their task will be great.

Enrolment in serious science courses is falling off in the English-speaking world. The reason for this seems to be that the courses are hard and demand more attention than many students consider worthwhile. It is difficult to know what percentage of the race (outside possible genetic manipulation in the future) are intended for the intellectual difficulty of a modern scientific education. Even the mathematical demands on lower order technicians may be greater than can be met. In a society where it is easy to earn one's living, why should people drive themselves to the pressure of such an education? But our society requires that more and more of its members be kept in educational institutions for longer and longer periods. Popularized humanities, handmaidens of the performing arts, will provide, along with a simplified sociology, the education easy enough to occupy the time of many.

Beyond this immediacy, however, lies a nobler reason for the care about art and the humanities in this era. When truth in science seems to teach us that we are accidental inhabitants of a negligible planet in the endless spaces, men are forced to seek meaning in other ways than through the intellect. If truth leads to meaninglessness, then men in their thirst for meaning turn to art. To hope to find in the products of the imagination that meaning which has been cast out of the intellect may, in the light of Socrates, be known to be a fruitless quest. Nevertheless, it is a thirst which is the enemy of tyranny.

IF WE ARE TO LIVE in the modern university as free men, we must make judgements about the essence of the university — its curriculum. If such judgements are to be more than quibbles about detail, they must be based on what we think human life to be, what activities serve human fulfilment, and what place higher education should play in encouraging the realization of these activities.

As soon as this is said, however, the tightness of the circle in which men find themselves in modern civilization becomes evident. For on one side of the picture, most men have given up not only the two great accounts of human excellence in the light of which western men had understood the purpose of existence (the one given in philosophy, the other in revealed religion), but also the very idea of human existence having a given highest purpose, and therefore an excellence which could be known and in terms of which all our activities could be brought into some order. It is now generally assumed that the race has meaning (call it if you will purpose) only on the condition that we view ourselves as purposive and that none of these views are truths concerning the nature of things, but only ideologies which we create to justify our man-made purposes. There is no objective purpose to human or non-human nature which men can come to know and in terms of which the various occasions of life can be ordered. Purpose and value are the creations of human will in an essentially purposeless world.

Yet it is not simply this absence of the idea of objective human excellence which constitutes the tightness of the circle in which we live. For on the other side of the picture, it is not to be thought that just because the dominant intellectual position of the age is that there is no highest purpose, the public realm is in fact able to do without such a conception. The political aspect of the liberal criticism of human excellence was the belief that unfettered by "dogmatic" and "a priori" ideas of excellence men would be free to make the world according to their own values and each would be able to fulfil his individuality. Ideas of purpose and indeed of a highest activity were superstitious strait-jackets — not only the enemies of an objective science, but of the free play of individuality. Their elimination through criticism was the first step towards building a pluralist society. Yet pluralism has not been the result in those societies where modern liberalism has prevailed. Western men live in a society the public realm of which is dominated by a monolithic certainty about excellence — namely that the pursuit of technological efficiency is the chief purpose for which the community exists. When modern liberals, positivist or existentialist, have criticized the idea of human excellence, they may have thought that they were clearing the ground of religious and metaphysical superstitions which stood in the way of the liberty of the individual. Rather they were serving the social purpose of legitimizing the totally technological society by destroying anything from before the age of progress which might inhibit its victory. Modern liberalism has been a superb legitimizing instrument for the technological society, because at one and the same time it has been able to criticize out of the popular mind the general idea of human excellence and yet put no barrier in the way of that particular idea of excellence which in fact determines the actions of the most powerful in our society. The mark of education is claimed to be scepticism about the highest human purposes, but in fact there is no scepticism in the public realm about what is important to do.

The fact that in our society the demands of technology are themselves the dominating morality is often obscured by the fact that the modern scientific movement has been intimately associated with the moral striving for equality. Mastery over the world would enable men to build a society in which all members would be freed from the tyranny of labour and for the benefits of leisure — the greatest of which was education. Leisure is only possible with the division of labour. But the division of labour without modern mastery resulted in inequality — particularly the grossest inequality in leisure. The noblest expectation of the age of progress was to overcome that limitation by building a society in which all men would come to have leisure through the mastery that science would make possible.[6] In the egalitarian faith it was believed that with

---

6 It is here that the profound connection can be seen between the age of progress and western Biblical religion — Judaism and Christianity. The centre of these religions lay in revelation which was a kind of knowledge concerning the most important matters, and which made all men in some sense equal by their potential openness to it. It was similar to the philosophical contemplation of the classical philosophers in that it dealt with the most important matters; it was different in that openness to it depended on faith. Can the

leisure all men would be open to philosophy and science, and so be able to choose rational goals for their own lives and for their communities. With such a hope (never more unequivocally present than in nineteenth century America) the problem of public education became pressing. This great hope has sustained this continent morally until recently, and is still often used as the rhetoric to obscure the emptiness of the curriculum to all but the claims of mastery.

I cannot describe here the complex history of how the progressive hope in American education was gradually emptied of all content except means to technological regulation and expansion. It is a long and complex road from the liberal Protestant believers of Massachusetts to the end of ideology. The rift in the lute can be seen early from the fact that as vigorous an exponent of the American morality as William James could say: "Yes, yes, we must have large things first, size first; the rest will come." Men such as Dewey had been profoundly influenced by Rousseau and his desire to give educational content to the life of the equal citizen; unlike Rousseau, they had an unlimited faith in the conquest of nature as the means to a more than formal equality. They did not pay attention to the side of Rousseau's thought in which he asserted that progress in the arts and sciences inhibited egalitarian virtue. The demands of the increasingly complex apparatus have dissipated the dream of modern liberalism which sought through education to give substance to equality. The mastery once thought of as a means becomes increasingly the public end.

To take a recent example: the optimistic Freudianism, so popular a faith among the enlightened, understood the central means to maturity to be the coming to terms with the history of gratification of our various orifices. Whatever the incompleteness of this Rousseauian Freudianism as an account of human excellence, it is clear that it claims to be a doctrine of individual and social happiness which must be achieved in a more than technical way. Its proponents such as Erikson show it as more than simply the servant of mastery. It is, however, a far cry from

---

modern belief in equality be understood apart from the change of emphasis (concerning man's highest activity) from contemplation to charity, which came with the dominance of Christianity? As has so often been said, the beliefs of the age of progress are a form of secularized Christianity.

such sexual humanism to the behaviourist psychology which now dominates our universities and which is geared to produce regulators who will fit the masses into the system by largely mechanical means. The humane if limited purposes of American Freudianism were not appropriate to the immensity of the institutions necessary to the over-coming of chance. The pressing need to regulate and control those institutions meant that the means of Freudian psychotherapy were too inward and individual, too chancy and too expensive to be used for any more than the privileged few. What was required, and what will be forthcoming, are immediately applicable techniques for the control of human nature on a grand scale. This requires the turning of the academic study of psychology to its present behaviourism.

The tight circle then in which we live is this: our present forms of existence have sapped the ability to think about standards of excellence and yet at the same time have imposed on us a standard in terms of which the human good is monolithically asserted. Thus, the university curriculum, by the very studies it incorporates, guarantees that there should be no serious criticism of itself or of the society it is shaped to serve. We are unable seriously to judge the university without judging its essence, the curriculum; but since we are educated in terms of that curriculum it is guaranteed that most of us will judge it as good. The criteria by which we could judge it as inadequate in principle can only be reached by those who through some chance have moved outside the society by memory or by thought. But so to have moved means that one's criticisms will not be taken seriously from within the society.[7]

It would be presumptuous to end by proposing some particular therapy by which we might escape from the tight circle of the modern fate. The decisions of western men over many centuries have made our world too ineluctably what it is for there to be any facile exit. Those who

---

7 This is not to deny that there will be all kinds of criticism of the university from within itself. How it should be governed, how the students should be treated as persons, whether research is sweeping away good teaching? etc., etc. Such criticisms are immensely welcomed because they serve as evidence that the society is still free and forward moving. The President of the University of Toronto praised the Macpherson report as "radical."

by some elusive chance have broken with the monolith will return to the problem of human excellence in ways too various to be procrusteanly catalogued. The sheer aridity of the public world will indeed drive many to seek excellence in strange and dangerous kingdoms (as those of drugs and myth and sexuality). In such kingdoms, moderation and courage may be known by the wise to be essential virtues. But when such virtues have been publicly lost they cannot be inculcated by incantation, but only rediscovered in the heat of life where many sparrows fall. Much suffering will be incurred by those who with noble intent follow false trails. Who is to recount how and when and where private anguish and public catastrophe may lead men to renew their vision of excellence?

In the realm of the academic, one of the essential therapies will be the reliving of buried memories of what the greatest, whether western or eastern, have known of human excellence. This rediscovery of the past will not be accomplished by those who view it as the task simply of technical scholarship, unrelated to what we are now; but by those who in many aspects of their lives, political, sexual, religious, etc., seek in the past the truth which they have here found wanting. Nor will such search be confined to particular disciplines, the specialists of which see this past as their private preserve. All sorts and conditions of students will find in a multitude of subjects means to transcend the aridity of the technological tradition. These means may be realized most openly and nobly by those who spend their lives in the most modern studies. Philosophy may be regained by those immersed in understanding the immediacies of the public world; reverence rediscovered in psychiatric researches.

It is possible, nevertheless, to assert one criterion by which all the potential therapies may be judged. Do they mitigate the division which comes forth from the modern vision? Do they help to overcome the way that we envision ourselves as "creative" freedom and all else as objects either useful, threatening, or indifferent? In its political context that division has led us, in our very drive to universalize freedom, to build the acme of the objective society which increasingly stifles the spontaneity of those it was built to free. The division widens so that it has almost killed what little remains of those mediators — common sense, reverence, communities, and art (perhaps even finally sexuality) — which are the means for us to cross the division separating ourselves and our habitations. At any time or place it is a strange destiny to be a "rational animal" — and indeed strange that there should be such — but the loss of these mediators makes that strangeness almost unbearable by tearing apart that which we are — rational animals. Socrates' prayer for the unity of the inward and the outward was spoken in an antique world, the context of which it could not be our historical business to recreate. Yet the fact begins to appear through the modernity which has denied it: human excellence cannot be appropriated by those who think of it as sustained simply in the human will, but only by those who have glimpsed that it is sustained by all that is. Although that sustainment cannot be adequately thought by us because of the fragmentation and complexity of our historical inheritance, this is still no reason not to open ourselves to all those occasions in which the reality of that sustaining makes itself present to us.

A Platitude

WE CAN HOLD IN OUR MINDS the enormous benefits of technological society, but we cannot so easily hold the ways it may have deprived us, because technique is ourselves. All descriptions or definitions of technique which place it outside ourselves hide from us what it is. This applies to the simplest accounts which describe technological advance as new machines and inventions as well as to the more sophisticated which include within their understanding the whole hierarchy of interdependent organizations and their methods. Technique comes forth from and is sustained in our vision of ourselves as creative freedom, making ourselves, and conquering the chances of an indifferent world.

It is difficult to think whether we are deprived of anything essential to our happiness, just because the coming to be of the technological society has stripped us above all of the very systems of meaning which disclosed the highest purposes of man, and in terms of which, therefore, we could judge whether an absence of something was in fact a deprival. Our vision of ourselves as freedom in an indifferent world could only have arisen in so far as we had analysed to disintegration those systems of meaning, given in myth, philosophy, and revelation, which had held sway over our progenitors. For those systems of meaning all mitigated both our freedom and the indifference of the world, and in so doing put limits of one kind or another on our interference with chance and the possibilities of its conquest.

It may be said that to use the language of deprival is to prejudice the issue, because what has gone can more properly be described as illusions, horizons, superstitions, taboos which bound men from taking their fate into their own hands. This may be the case. What we lost may have been bad for men. But this does not change the fact that something has been lost. Call them what you will — superstitions or systems of meaning, taboos or sacred restraints — it is true that most western men have been deprived of them.

It might also be said that the older systems of meanings have simply been replaced by a new one. The enchantment of our souls by myth, philosophy, or revelation has been replaced by a more immediate meaning — the building of the society of free and equal men by the overcoming of chance. For it is clear that the systematic interference with chance was not simply undertaken for its own sake but for the realization of freedom. Indeed it was undertaken partly in the name of that charity which was held as the height in one of those ancient systems of meaning. The fulfillment that many find in the exploration of space is some evidence that the spirit of conquest may now be liberated from any purpose beyond itself, since such exploration bears no relation to the building of freedom and equality here on earth. What we are can be seen in the degree to which the celebration of the accomplishments in space is not so much directed to the value of what has actually been done, but rather to the way this serves as verification of the continuing meaning in the modern drive to the future, and the possibility of noble deeds within that drive. Be that as it may, the building of the universal and homogeneous state is not in itself a system of meaning in the sense that the older ones were. Even in its realization, people would still be left with a question, unanswerable in its own terms: how do we know what is worth doing with our freedom? In myth, philosophy, and revelation, orders were proclaimed in terms of which freedom was measured and defined. As freedom is the highest term in the modern language, it can no longer be so enfolded. There is therefore no possibility of answering the question: freedom for what purposes? Such may indeed be the true account of the human situation: an unlimited freedom to make the world as we want in a universe indifferent to what purposes we choose. But if our situation is such, then we do not have a system of meaning.

All coherent languages beyond those which serve the drive to unlimited freedom through technique have been broken up in the coming to be of what we are. Therefore it is impossible to articulate publicly any suggestion of loss, and perhaps even more frightening, almost impossible to articulate it to ourselves. We have been left with no words which cleave together and summon out of uncertainty the good of which we may sense the dispossession. The drive to the planetary technical future is in any case inevitable; but those who would try to divert, to limit, or even simply to stand in fear before some of its applications find themselves defenceless, because of the disappearance of any speech by which the continual changes involved in that drive could ever be thought as deprivals. Every development of technique is an exercise of freedom by those who develop it, and as the exercise of freedom is the only meaning, the changes can only be publicly known as the unfolding of meaning.

I am not speaking of those temporary deficiencies which we could overcome by better calculations — e.g., cleverer urbanologists — failures of the system which may be corrected in its own terms. Nor do I mean those recognitions of deprivation from the dispossessed — either amongst us or in the southern hemisphere — who are conscious of what they have not got and believe their lack can be overcome by the humane extension of the modern system.

Also, in listening for the intimations of deprival either in ourselves or others we must strain to distinguish between differing notes: those accidental deprivals which tell us only of the distortions of our own psychic and social histories, and those which suggest the loss of some good which is necessary to man as man. As I have said elsewhere, thought is not the servant of psychoanalysis or sociology; but a straining for purification has the authority of the Delphic "know thyself," and of the fact that for Socrates the opposite of knowledge was madness. The darkness of the rational animal requires therapy, and now that "philosophy" sees itself as analysis, men who desire to think must include in their thinking those modern therapies which arose outside any connection with what was once called philosophy. This inclusion of what may be health-giving in psychoanalysis and sociology will be necessary, even within the knowledge that these therapies are going to be used unbridledly as servants of the modern belief that socially useful patterns of behaviour should be inculcated by force. Anything concerning sexuality will serve as an example of the distortions I am trying to describe, because such matters touch every element of fantasy and the unconscious in ourselves, so that judgements about good are there most clouded by idiosyncracy. For example, if a man were to say that the present technical advances were so detaching sexuality from procreation as to deprive women of a maternity necessary to their fullest being, his statement might be suspect as coming from a hatred of women which could not bear to see them free. To take an example from myself, a sophisticated and lucid sociologist has asserted that I was saddened at the disappearance of the Canadian nation into the American empire, not because of my written reasons from political philosophy but because of my biographically determined situation.[1] I belonged to a class which had its place in the old Canada and could find no place for itself in the new imperial structure. Or again, I know that my thinking about modern liberalism is touched by a certain animus arising from tortured instincts, because of the gynarchy in which I came to know that liberalism. Thought may first arise from the ambiguities of personal history but if it is to stand fairly before the enormous ambiguities of the dynamo, it must attempt to transcend the recurring distortions of personal history. To listen for the intimations of deprival requires attempting a distinction between our individual history and any account which might be possible of what belongs to man as man.

Yet even as one says this, the words fade. The language of what belongs to man as man has long since been disintegrated. Have we not been told that to speak of what belongs to man as man is to forget that man creates himself in history? How can we speak of excellences which define the height for man, because what one epoch calls an excellence another does not, and we can transcend such historical perspectives only in the quantifiable? Aren't such excellences just a crude way of talking about values, pretending that they have some status in the nature of things beyond our choosing? We are back where we began: all languages of good except the language of the drive to freedom have disintegrated, so it is just to pass some antique wind to speak of goods that belong to man as man. Yet the answer is also the same: if we cannot so speak, then we can either only celebrate or stand in silence before that drive. Only in listening for the intimations of deprival can we live critically in the dynamo.

---

1 R. K. Crook "Modernization and Nostalgia—a note on the sociology of pessimism," *Queens Quarterly*, 1965.

Whether there are intimations of essential deprivals which are beyond elimination by the calculations of the present spirit is just what must remain ambiguous for us, because the whole of our dominating system of thought denies that there could be such. When we sense their arising, at the same time we doubt that which we sense. But even among some of those who use the language of sheer freedom as protest, there seems to be heard something different than the words allow. Because they have been taught no language but the modern, they use it not only to insist that the promises of the modern be fulfilled, but also to express their anguish at its denials.

Any intimations of authentic deprival are precious, because they are the ways through which intimations of good, unthinkable in the public terms, may yet appear to us. The affirmation stands: how can we think deprivation unless the good which we lack is somehow remembered? To reverse the platitude, we are never more sure that air is good for animals than when we are gasping for breath. Some men who have thought deeply seem to deny this affirmation: but I have never found any who, in my understanding of them, have been able, through the length and breadth of their thought, to make the language of good secondary to freedom. It is for this reason that men find it difficult to take despair as the final stance in most circumstances. Deprivation can indeed become absolute for any of us under torture or pain or in certain madnesses. We can be so immersed in the deprival that we are nothing but deprivation. Be that as it may, if we make the affirmation that the language of good is inescapable under most circumstances, do we not have to think its content? The language of good is not then a dead language, but one that must, even in its present disintegration, be re-collected, even as we publicly let our freedom become ever more increasingly the pure will to will.

We know that this re-collection will take place in a world where only catastrophe can slow the unfolding of the potentialities of technique. We cannot know what those potentialities will be. I do not simply mean specific possibilities — for example, whether housework will be done by robots, how far we can get in space, how long we can extend the life span, how much we can eliminate socially undesirable passions, in what ways we can control the procreation of the race, etc., etc. Some of these possibilities we can predict quite clearly, others we cannot, or not yet.

But what is more important, we cannot know what these particular possibilities tell us about the potential in the human and the non-human. We do not know how unlimited are the potentialities of our drive to create ourselves and the world as we want it. For example, how far will the race be able to carry the divided state which characterises individuals in modernity: the plush patina of hectic subjectivity lived out in the iron maiden of an objectified world inhabited by increasingly objectifiable beings? When we are uncertain whether anything can mediate that division, how can we predict what men will do when the majority lives more fully in that division? Is there some force in man which will rage against such division: rage not only against a subjectivity which creates itself, but also against our own lives being so much at the disposal of the powerful objectifications of other freedoms? Neither can we know what this unfolding potentiality tells us of the non-human. As we cannot now know to what extent the non-human can in practice be made malleable to our will, therefore we also do not know what this undetermined degree of malleability will tell us of what the non-human is. Is the non-human simply stuff at our disposal, or will it begin to make its appearance to us as an order the purposes of which somehow resist our malleablizings? Are there already signs of revolts in nature?

Despite the noblest modern thought, which teaches always the exaltation of potentiality above all that is, has anyone been able to show us conclusively throughout a comprehensive account of both the human and non-human things, that we must discard the idea of a presence above which potentiality cannot be exalted? In such a situation of uncertainty, it would be lacking in courage to turn one's face to the wall, even if one can find no fulfillment in working for or celebrating the dynamo. Equally it would be immoderate and uncourageous and perhaps unwise to live in the midst of our present drive, merely working in it and celebrating it, and not also listening or watching or simply waiting for intimations of deprival which might lead us to see the beautiful as the image, in the world, of the good.

LIST

The A List

The Outlander Gil Adamson
The Circle Game Margaret Atwood
Power Politics Margaret Atwood
Second Words Margaret Atwood
Survival Margaret Atwood
These Festive Nights Marie-Claire Blais
La Guerre Trilogy Roch Carrier
The Hockey Sweater and Other Stories Roch Carrier
Hard Core Logo Nick Craine
Great Expectations Edited by Dede Crane and Lisa Moore
Queen Rat Lynn Crosbie
The Honeyman Festival Marian Engel
The Bush Garden Northrop Frye
Eleven Canadian Novelists Interviewed by Graeme Gibson
Five Legs Graeme Gibson
Death Goes Better with Coca-Cola Dave Godfrey
De Niro's Game Rawi Hage
Kamouraska Anne Hébert
Ticknor Sheila Heti
No Pain Like This Body Harold Sonny Ladoo
Civil Elegies Dennis Lee
Mermaids and Ikons Gwendolyn MacEwen
Ana Historic Daphne Marlatt
Like This Leo McKay Jr.
The Selected Short Fiction of Lisa Moore
Furious Erín Moure
Selected Poems Alden Nowlan
Poems for All the Annettes Al Purdy
Manual for Draft-Age Immigrants to Canada Mark Satin
The Little Girl Who Was Too Fond of Matches Gaétan Soucy
Made for Happiness Jean Vanier
Basic Black with Pearls Helen Weinzweig
Passing Ceremony Helen Weinzweig
The Big Why Michael Winter
This All Happened Michael Winter